The (mis)Behavior of Markets

ALSO BY BENOIT B. MANDELBROT

Les objets fractals: forme, hasard et dimension
(1975, 1984, 1989, 1995)

Fractals: Form, Chance and Dimension
(1977)

The Fractal Geometry of Nature
(1982)

Fractals and Scaling in Finance:
Discontinuity, Concentration, Risk
(1997)

Fractales, hasard et finance (1959–1997)
(1997)

Multifracals and 1/f Noise: Wild Self-Affinity in Physics
(1999)

Gaussian Self-Affinity and Fractals:
Globality, the Earth, 1/f, and R/S
(2002)

Fractals, Graphics, and Mathematics Education
(With M. L. Frame)
(2002)

Fractals and Chaos:
The Mandelbrot Set and Beyond
(2004)

The (mis)Behavior of Markets

. . . .

A Fractal View of Risk, Ruin, and Reward

Benoit B. Mandelbrot and Richard L. Hudson

BASIC
BOOKS

A MEMBER OF THE PERSEUS BOOKS GROUP

NEW YORK

Hardcover first published in 2004 by Basic Books,
A Member of the Perseus Books Group
Paperback first published in 2006 by Basic Books

Books published by Basic Books are available at special discounts for bulk
purchases in the United States by corporations, institutions, and other
organizations. For more information, please contact the Special Markets
Department at the Perseus Books Group, 11 Cambridge Center, Cambridge,
MA 02142, or call (617) 252-5298, (800) 255-1514 or e-mail
special.markets@perseusbooks.com.

Library of Congress Cataloging-in-Publication Data
Mandelbrot, Benoit B.
The (mis)behavior of markets : a fractal view of risk, ruin, and reward /
Benoit B. Mandelbrot and Richard L. Hudson.
p. cm.
Includes bibliographical references and index.
HC: ISBN-13 978-0-465-04355-2; ISBN 0-465-04355-0 (alk. paper)
1. Capital market. 2. Investment analysis. 3. Stocks—Prices. 4.
Securities. 5. Risk management. I. Title: Misbehavior of markets. II.
Title: Behavior of markets. III. Hudson, Richard L. IV. Title.
HG4523.M257 2004
332'.01—dc22
2004011400

Book design by Lovedog Studio

PBK: ISBN-13 978-0-465-04357-6; 0-465-04357-7

10 9 8 7 6 5 4 3 2

To the Scientific Reader: An Abstract

Three states of matter—solid, liquid, and gas—have long been known. An analogous distinction between three states of randomness—mild, slow, and wild—arises from the mathematics of fractal geometry. Conventional financial theory assumes that variation of prices can be modeled by random processes that, in effect, follow the simplest "mild" pattern, as if each uptick or downtick were determined by the toss of a coin. What fractals show, and this book describes, is that by that standard, real prices "misbehave" very badly. A more accurate, multifractal model of wild price variation paves the way for a new, more reliable type of financial theory.

Understanding fractally wild randomness, also exemplified by such diverse phenomena as turbulent flow, electrical "flicker" noise, and the track of a stock or bond price, will not bring personal wealth. But the fractal view of the market is alone in facing the high odds of catastrophic price changes. This book presents this view in a highly personal style, with many pictures and no mathematical formula in the main text.

Dedication

*Aux Dames: Aliette, Diane, Louisa,
Clara et Ruth*

Contents

> *Benoit Mandelbrot, the "father" of fractals, has made a career of going against the prevailing fashions in science.*

Part I.
The Old Way

> *"Modern" financial theory is founded on a few, shaky myths that lead us to underestimate the real risk of financial markets.*

> *How the operations of mere chance can be used to study a financial market.*

> *The study of financial theory began a century ago with a brilliant but undervalued French mathematician, Louis Bachelier.*

Part II.
The New Way

> *The second clue to fractal finance came from lifelong study of
> the Nile River by an English hydrologist, H.E. Hurst.*

> *The two critical features of financial markets are wild price
> swings and long-term dependence—the Noah Effect and the
> Joseph Effect.*

> *In financial markets, time speeds up and slows down—as
> described in the multifractal model of markets.*

Part III.
The Way Ahead

> *How do financial markets really work? A list of key insights
> provided by the fractal view of finance.*

> *So how can the study of fractals change finance? A program for
> future research.*

Acknowledgments

NO BOOK IS MADE ALONE. In this instance the help and support of many people have been essential. Here they are acknowledged with gratitude.

Survival when taking high risks is often a reward for good timing. This is how Professor Mandelbrot repeatedly escaped ruin on his way to fractals. He is deeply in debt to the Thomas J. Watson Research Center of IBM—for thirty-five years a unique haven for mavericks engaged in investigations that science and society deemed desirable but had few ways of supporting. To list every helpful colleague would be impossible; but worthy of special mention is Ralph E. Gomory, to whom Mandelbrot was fortunate to report in various ways for much of his time at IBM. Upon retirement, Mandelbrot was brought to the Yale Mathematics Department by Ronald R. Coifman and Peter W. Jones, who opened to him another exceptional haven. Throughout, Aliette K. Mandelbrot provided extremely active participation, excellent advice, and unfailing enthusiasm.

For his part, Mr. Hudson would like to thank those who have encouraged his own small forays into risk, whether professional or personal. At Katholieke Universiteit Leuven, in Belgium, Dean Filip Abraham and Professor Paul De Grauwe of the Faculty of Economics and Applied Economics provided vital support and friendship with their offer for Mr. Hudson to work on this book as a visiting scholar in their midst. At the *Wall Street Journal*, Frederick S. Kempe encouraged this enterprise as both colleague and friend, and Paul E. Steiger and Karen Elliott House graciously granted leave to undertake it. And at home, Diane M. Fresquez was a guiding spirit. She helped review and research portions of the book; patiently transcribed many hours of tape-recorded discussions between the authors; and provided—as ever—her generous encouragement and wise companionship.

For the art, we thank M. Gruskin, H. Kanzer, and M. Logan.

P R E L U D E
by Richard L. Hudson

Introducing a Maverick in Science

INDEPENDENCE IS A GREAT VIRTUE. To illustrate that, Benoit Mandelbrot relates how, during the German occupation of France in World War II, his father escaped death. One day, a band of Resistance fighters attacked the prison camp where he was being held. They disarmed the guards and told the inmates to flee before the main German force struck back. So the surprised and disoriented prisoners set off towards nearby Limoges, en masse and on the high road. After half a kilometer, Mandelbrot *père* decided this way was folly. So he set off by himself. He left the main group and the open road and broke off into the thick forest to walk back home alone. Shortly after, he heard a German Stuka dive-bomber strafe the main party of prisoners on the high road. He, alone in the forest, escaped harm. "It was," recalls the son, "the way my father behaved throughout his life. He was an independent man—and so am I."

Mandelbrot, a teenager during the war, is now famous. He got a Ph.D. in mathematical sciences in Paris, joined the influx of European scientists to America, and went on to a long career of scientific discovery and acclaim. He invented a new branch of mathematics, fractal geometry; he applied it to dozens of improbably diverse fields; and he received numerous awards and much media attention. But his early wartime lessons in independence—he says he was *aguerri*, or war-hardened, by his experiences—made him always strike off in a direction different from the rest. He has thereby engendered much controversy, through which he persisted. He calls himself a maverick. By that, he means he has spent his life doing only what he felt right, sticking his nose where it was not always wanted, belonging to no particular scientific community.

"I have been a lone rider so often and for so long, that I'm not even bothered by it anymore," he says. Or, as a mathematically minded friend put it, he moves orthogonally—at right angles—to every fashion.

These facts about Mandelbrot's life are important to remember when meeting him, as in this book. What he says is not what they normally teach at the business schools at Harvard, London, Fontainebleau, or his own university, Yale. He has been premature, contrary to fashion, trouble-making, in virtually every field he has touched: statistical physics, cosmology, meteorology, hydrology, geomorphology, anatomy, taxonomy, neurology, linguistics, information technology, computer graphics, and, of course, mathematics. In economics he is especially controversial. His first appearance in the field, in the early 1960s, caused a storm. Paul H. Cootner, then a well-known economist at MIT, praised Mandelbrot's work as "the most revolutionary development in the theory of speculative prices" since the study began in 1900—and then he went on to criticize details of its contents and "Messianic tone." It has been like that ever since. The economics establishment knows him well, finds him intriguing, and has grudgingly adopted many of his ideas (though often without giving him full credit). That has made him one of the most important forces for change in the theory of finance. But the establishment also finds him bewildering.

So this book is an end-run, to a broader world and a broader audience than can be found in the faculty lounges of Cambridge, Massachusetts, or Cambridge, England. What Mandelbrot has to say is important and immediately relevant to every professional in finance, every investor in the market, anyone who just wants to understand how money gets won and lost with such frightening rapidity.

From the start, Mandelbrot has approached the market as a scientist, both experimental and theoretical. Einstein famously said: "The grand aim of all science is to cover the greatest number of empirical facts by logical deduction from the smallest number of hypotheses or axioms." Such parsimony has been Mandelbrot's aim. To him, a stock exchange is a "black box," a system at once complex, variegated, and elusive, to be studied with conceptual and mathematical tools that build upon those of physics. Since he pioneered this approach in the 1960s, it has greatly evolved. It provides a scientific perspective on markets that is unlike any you will find in conventional books on investment, markets, and the economy.

Thus, reading this volume will not make you rich. But it will make you wiser—and may thereby save you from getting poorer.

I, CO-AUTHOR in this endeavor, first met Mandelbrot in 1997 when I was managing editor of the *Wall Street Journal*'s European edition. He showed up at our Brussels office with a mission to convince us that we should rethink how markets work. At first, he struck me as the "mad scientist" stereotype—flyaway white hair, very cerebral, intense convictions, a fondness for digression and disputation. But I and editor and publisher Phil Revzin, then my boss, listened politely and did what newspaper editors often do in such circumstances. What the heck? Print what he has to say, and see what happens.

A year later, when I was planning a business conference for the newspaper, I thought of inviting Mandelbrot to talk about risk. He stole the show. The conference-goers, among the best-known financiers, entrepreneurs, and CEOs in Europe—preeminent risk-takers, all—listened at first in bemusement. Not your usual conference speaker. Then they got sucked into his strange story. Some said he made more sense than their CFOs. Afterwards, in our audience-feedback survey, they rated him as best speaker of the day—tied only by Steve Ballmer, the Microsoft CEO.

As a scientist, Mandelbrot's fame rests on his founding of fractal geometry, and on his showing how it applies in many fields. A fractal, a term he coined from the Latin for "broken," is a geometric shape that can be broken into smaller parts, each a small-scale echo of the whole. The branches of a tree, the florets of a cauliflower, the bifurcations of a river—all are examples of natural fractals. The math eschews the smooth lines and planes of the Greek geometry we learn in school, but it has astonishingly broad applications wherever roughness is present—that is, nearly everywhere. Roughness is the central theme of his work. We have long had precise measurements and elaborate physical theories for such basic sensations as heat, sound, color, and motion. Until Mandelbrot, we never had a proper theory of the irregular, the rough—all the annoying imperfections that we normally try to ignore in life. Roughness is in the jagged edge of a metal fracture, the rugged coastline of Britain, the static on a phone line, the gusts of the wind—even the irregular charts of a stock index or exchange rate. As he puts it, "Roughness is the uncontrolled element in life."

Studying roughness, Mandelbrot found fractal order where others had only seen troublesome disorder. His manifesto, *The Fractal Geometry of Nature*, appeared in 1982 and became a scientific best-seller. Soon, T-shirts and posters of his most famous fractal creation, the bulbous but infinitely complicated Mandelbrot Set, were being made by the thousands. His ideas were also embraced immediately by another scientific movement, chaos theory. "Fractals" and "chaos" entered the popular vocabulary. In 1993, on receiving the prestigious Wolf Prize for Physics, Mandelbrot was cited for "having changed our view of nature."

MANDELBROT'S LIFE story has been a tale of roughness, irregularity,. and what he calls "wild" chance. He was born in Warsaw in 1924, and tutored privately by an uncle who despised rote learning; to this day, Mandelbrot says, the alphabet and times tables trouble him mildly. Instead, he spent most of his time playing chess, reading maps, and learning how to open his mind to the world around him.

His harsh education in war came soon enough. Unusually attentive to the footsteps of approaching trouble, the Jewish family moved in 1936 to Paris, where another uncle, Szolem Mandelbrojt (spellings differ in so wandering a family), had settled earlier as a mathematics professor. The war came, and young Mandelbrot was sent to a small town in the French countryside, at different times caring for horses or mending tools. An overcoat nearly undid him. His father had bought him a woolen coat in an orange, pseudo-Scotch plaid: It was hideous by anybody's standards, but warm and welcome in wartime. One day, the police stopped him and his younger brother. A tall man wearing just such an overcoat had been spotted earlier, fleeing the scene of a French Resistance attack on German headquarters. "That's him," a collaborator pointed. A case of mistaken identity. Mandelbrot was released, but took no chances: An opportunity arose, and he slipped out of town.

Mandelbrot's moment of self-discovery as a mathematician came in Lyon in 1944, where benefactors hid him in—appropriately—a school. He had a fake ID card and touched-up ration coupons. The staff asked no questions; theirs was, he recalls, "a passive kind of *résistance.*" In the first week, he sat uncomprehending before the meaningless words and numbers on the blackboard. Then the professor embarked on a lengthy algebraic journey. Mandelbrot's hand shot up. "Sir, you don't need to make any calculations. The answer is obvious." He described a geometrical approach that yielded a fast, simple solution. Where others would have used a formula, he saw a picture. The teacher, skeptical at first, checked: Correct. And

Mandelbrot kept doing the same thing, in problem after problem, in class after class. As he relates it:

> It happened so fast I was not conscious of it. I would say to myself: This construction is ugly, let's make it nicer. Let's make it symmetric. Let's project it. Let's embed it. And all that, I could see in perfect 3-D vision. Lines, planes, complicated shapes.

Ever since, pictures have been his special aids to inspiration and communication. Some of his most important insights came, not from elaborate mathematical reasoning, but from a flash recognition of kinship between disparate images—the strange resemblance between diagrams concerning income distribution and cotton prices, between a graph of wind energy and of a financial chart. The creative essence of fractal geometry is to combine the formal and the visual. The ready intuition of fractal pictures has, today, made the subject a college course at Yale and other universities, and a popular addition to many high school math courses. But among "pure" mathematicians, Mandelbrot's approach was initially criticized. Not rigorous, they chided; the eye can mislead. But, Mandelbrot rejoins, observation often led him to conjectures that have stimulated and challenged the most skilled mathematicians; many of these problems remain unsolved. In any event, when science was young, he says, pictures were essential; think of the anatomical drawings of Vesalius, the engineering sketches of Leonardo, or the optics diagrams of Newton. Only in the nineteenth century, when the great edifice of algebraic analysis was perfected, did pictures become suspect as, somehow, imprecise.

In an ever-more complex world, Mandelbrot argues, scientists need both tools: image as well as number, the geometric view as well as the analytic. The two should work together. Visual geometry is like an experienced doctor's savvy in reading a patient's complexion, charts, and X-rays. Precise analysis is like the medical test results— the raw numbers of blood pressure and chemistry. "A good doctor looks at both, the pictures and the numbers. Science needs to work that way, too," he says.

Mandelbrot's career has taken a jagged path. In 1945, he dropped out of France's most prestigious school, the École Normale Supérieure, on the second day, to enroll at the less-exalted but more appropriate École Polytechnique. He proceeded to Caltech; then—after a Ph.D. in Paris—to MIT; then to the Institute for Advanced Study in Princeton, as the last post-doc to study with the great Hungarian-born mathematician, John von Neumann; then to Geneva and back to Paris for a time.

Atypically for a scientist in those days, Mandelbrot ended up working, not in a university lecture hall, but in an industrial laboratory, IBM Research, up the Hudson River from Manhattan. At that time IBM's bosses were drawing into that lab and its branches a number of brainy, unpredictable people, not doubting they would do something brilliant for the company. In all kinds of ways, it was a wise policy. Scientifically, it yielded five Nobel Prize winners. But it was abandoned in the 1990s, as the company struggled to survive. Mandelbrot's research for IBM included the patterns of errors in computer communication and applications of computer analysis—even, at one point, for the company's president an investigation of stock-price behavior. During the 1980s, his computer-drawn Mandelbrot Set became an oft-repeated demonstration and a test of the processing power of IBM's then-new personal computers. But Mandelbrot's scientific activities and reputation went far beyond the confines of the lab at Yorktown Heights.

FOR MANDELBROT, economics has been both inspiration and curse. His study of financial charts in the 1960s helped stimulate his subsequent fractal theories in the 1970s and 1980s. He taught economics for a year at Harvard; and his first major paper in the field in 1962 (expanded and revised in 1963 and the next few years) was a study of cotton prices. In it, he presented substantial evidence against one of the fundamental assumptions of what became "modern" financial theory. At that time, the theory was beginning to be entrenched in university economics departments—and it would soon become orthodoxy on Wall Street. As Mandelbrot continued his fractal studies, he often returned to economics. Each time, he probed how markets work, how to develop a good economic model for them—and, ultimately, how to avoid loss in them.

Today, some of his ideas are accepted as orthodoxy. As the last chapter will show, they are incorporated into some of the most-sophisticated mathematical models with which banks and brokerage houses manage money, into the ways math Ph.D.'s price exotic options or measure portfolio risk from Wall Street to the City of London. For the sake of historical precision, a technical listing is in order here. Mandelbrot was the first to take seriously and study the so-called power-law distributions. His 1962 argument that prices vary far more than the standard model allows—that their distributions have "fat tails"—is now widely accepted by econometricians. (Scientific nomenclature is not always straightforward. The probability distribution behind this particular approach is variously called L-stable, stable Paretian, Lévy, or Lévy-Mandelbrot.) Also accepted is his argument that, by their very essence, prices can vary by leaps and bounds rather than in a continuous blur; and likewise, his 1965 argument that price changes today are dependent on changes in the long past.

These are all facts of financial life that Mandelbrot established early on and insisted upon, even though they ran counter to the theology of finance that was becoming established at about the same time. He also did pioneering work in many now-well-trodden avenues of economics. From 1965 he was publishing on what he soon called fractional Brownian motion and on the underlying concept of fractional integration, which has recently become a widespread econometric technique. In 1972, he published a multifractal model that incorporates and extends long tails and long dependence. His papers from the 1960s are the pillars upon which rest a branch of the dismal science called "econophysics." In 1966 he developed a mathematical model explaining how rational market mechanisms can generate price "bubbles." And finally, he built multifractals on his 1967 notion of a "subordinated" trading time, developed with H. M. Taylor, that has also passed into the toolkit of some financial modelers—though it, like some of his other theories, is often credited to later researchers.

Indeed, as a financial journalist previously unmired in disputes of academic priority, I would say Mandelbrot's batting average for correctly analyzing market behavior would accord him a place in the Economics Hall of Fame. That record, alone, should make this book worth reading.

But plenty of Mandelbrot's other ideas remain controversial in economics: for instance, his theories of "scaling," of multifractal analysis, and of long-term dependence—all at the core of this book. One reason was hinted at in Cootner's original review. Before resuming his sharp-tongued critique, the MIT economist summarized the significance of what Mandelbrot had, at that early date, only begun to say:

> Mandelbrot, like Prime Minister Churchill before him, promises us not utopia but blood, sweat, toil and tears. If he is right, almost all of our statistical tools are obsolete—least squares, spectral analysis, workable maximum-likelihood solutions, all our established sample theory, closed distributions. Almost without exception, past econometric work is meaningless.

IN 2004, in his eightieth year, Mandelbrot continues making trouble. He works the same full schedule—including weekends—as he always has. He continues publishing new research papers and books, lecturing at Yale, and traveling the world of scientific conferences to advance his views. Why not? After all, as he points out, Racine's most enduring play, *Athalie*; Verdi's greatest opera, *Falstaff*;

Wagner's *Ring Cycle*—all were written in the twilight of life, when the artist, after years of experience and experimentation, was at the height of his powers.

This book, too, is somewhat of an operatic performance—an interplay of voices, drama, and scenery. Throughout the main body of the book, the "I" voice is that of Mandelbrot, the ideas are his, and it is the drama of their discovery that motivates much of the text. The scenery is extensive and elaborate: Pictures, charts, and diagrams are key to understanding. And like the best operas, this book is written to be both engaging and popular. As the Notes and Bibliography suggest, a wealth of solid science and mathematics underpin our assertions—and the curious scientist or economist is welcome to consult those sources. All readers, of whatever background, are invited to visit the online addenda, www.misbehaviorofmarkets.com. It descends partly from a truly extraordinary Web site at http://classes.yale.edu/fractals/index.html created by Mandelbrot's Yale colleague, Professor Michael Frame, for their popular undergraduate course on fractals for non-science majors, Math 190.

Today, Mandelbrot's message is more timely than ever, after a turbulent decade of bull markets, currency crises, bear markets, and the repeated building and bursting of asset bubbles. Financial markets are very risky places. And hitherto our understanding of them has been laden by the elaborate mathematics of orthodox financial theory—with many misguided assumptions, mis-applied equations, and misleading conclusions. Financial markets are complicated, but they need not be made overly so. To repeat: The aim of science is parsimony. The goal of this book is simplicity.

PART ONE

· · · ·

The Old Way

Chorale: The computer "bug" as artist, opus 2. (Overleaf) Computer-generated art from Mandelbrot 1982. This design was created by a "bug" in a software program while I was investigating various fractal forms—and it nicely demonstrates the creative power of chance, in art, finance and life.

CHAPTER I

Risk, Ruin, and Reward

IN THE SUMMER OF 1998, the improbable happened.

On Wall Street, the historic bull market of the New Gay '90s was looking tired. There was no single, overwhelming problem—just a series of worries: recession in Japan, possible devaluation in China, and in Washington a president battling impeachment. Then came news that Russia, just two years earlier the world's hottest emerging market, was hitting a cash crunch. Western banks and debt-traders would suffer; a few, it later emerged, were already near ruin. So on August 4, the Dow Jones Industrial Average fell 3.5 percent. Three weeks later, as news from Moscow worsened, stocks fell again, by 4.4 percent. And then again, on August 31, by 6.8 percent. Other markets reeled: Bank bonds plummeted a third from their usual value against government bonds. The hammer blows were shocking—and for many investors, inexplicable. It was a panic, irrational and unpredictable; "the culmination of a meltdown," one analyst told the *Wall Street Journal*. It might, said another, "take a lifetime for investors to ever recoup some of those losses."

So much for conventional market wisdom. As we know now, the International Monetary Fund patched Russia, the Federal Reserve stabilized Wall Street, and the bull market ran another few years. In fact, by the conventional wisdom, August 1998 simply should never have happened; it was, according to the standard models of the financial industry, so improbable a sequence of events as to have been impossible. The standard theories, as taught in business schools around the world, would estimate the odds of that final, August 31, collapse at one in 20 million—an event that, if you traded daily for nearly 100,000 years, you would not expect to see even once. The odds of getting three such declines in the same month were even more minute: about one in 500 billion. Surely, August had been supremely bad luck, a freak accident, an "act of God" no one could have predicted. In the language of statistics, it was an "outlier" far, far, far from the normal expectation of stock trading.

Or was it? The seemingly improbable happens all the time in financial markets. A year earlier, the Dow had fallen 7.7 percent in one day. (Probability: one in 50 billion.) In July 2002, the index recorded three steep falls within seven trading days. (Probability: one in four trillion.) And on October 19, 1987, the worst day of trading in at least a century, the index fell 29.2 percent. The probability of that happening, based on the standard reckoning of financial theorists, was less than one in 10^{50}—odds so small they have no meaning. It is a number outside the scale of nature. You could span the powers of ten from the smallest subatomic particle to the breadth of the measurable universe—and still never meet such a number.

So what's new? Everyone knows: Financial markets are risky. But in the careful study of that concept, risk, lies knowledge of our world and hope of a quantitative control over it.

For more than a century, financiers and economists have been striving to analyze risk in capital markets, to explain it, to quantify it, and, ultimately, to profit from it. I believe that most of the theorists have been going down the wrong track. The odds of financial ruin in a free, global-market economy have been grossly underestimated. In this sense, the common man is wise in his prejudice that—especially after the collapse of the Internet bubble—markets are risky. But financial theorists are not so wise. Over the past century, they devised an intricate mathematical apparatus for appraising risk. It was adopted wholesale by Wall Street in the 1970s. The likes of Merrill Lynch, Goldman Sachs, and Morgan Stanley made it a part of intricate trading strategies. They tried tuning investment portfolios to different frequencies of risk and reward, as one might tune a radio. But the financial bumps and lurches of the 1980s and 1990s have forced a rethink, among financiers as well as among economists. Black Monday of 1987, the Asian economic crisis of 1997, the Russian summer of 1998, and the bear market of 2001 to 2003—surely, many now realize, something is not right. If reward and risk make a ratio, the standard arithmetic must be wrong. The denominator, risk, is bigger than generally acknowledged; and so the outcome is bound to disappoint. Better assessment of that risk, and better understanding of how risk drives markets, is a goal of much of my work.

My life has been a study of risk. I learned about it firsthand in the brutal school of World War II, as a Polish refugee hiding in the French countryside with a borrowed identity and touched-up ration coupons, masquerading (badly) as a simple country boy in an occupied land. I faced it in my career, rejecting the safety of French academia for the intellectual wanderings of an industrial scientist in a more free-wheeling America. As a scientist, all of my research has, in one way or another, veered between the two poles of human experience: deterministic systems of order and planning, and stochastic, or random, systems of irregularity and unpredictability. My key contribution was to found a new branch of mathematics that perceives the hidden order in the seemingly disordered, the plan in the unplanned, the regular pattern in the irregularity and roughness of nature. This mathematics, called fractal geometry, has much to say in the natural sciences. It has helped model the weather, study river flows, analyze brainwaves and seismic tremors, and understand the distribution of galaxies. It was immediately embraced as an essential mathematical tool in the 1980s by "chaos" theory, the study of order in the seeming-chaos of a whirlpool or a hurricane. It is routinely used today in the realm of man-made structures, to measure Internet traffic, compress computer files, and make movies. It was the mathematical engine behind the computer animation in the movie, *Star Trek II: the Wrath of Khan*.

I believe it has much to contribute to finance, too. For forty years in fits and starts, as allowed by my personal interests, by unfolding events, and by the availability of colleagues to talk to, the development of fractal geometry has continually interacted with my studies of financial markets and economic systems. I have investigated them not as an economist or financier, but as a mathematical and experimental scientist. To me, all the power and wealth of the New York Stock Exchange or a London currency-dealing room are abstract; they are analogous to physical systems of turbulence in a sunspot or eddies in a river. They can be analyzed with the tools science already has, and new tools I keep adding to the old ones as need and ability allow. With these tools, I have analyzed how income gets distributed in a society, how stock-market bubbles form and pop, how company size and industrial concentration vary, and how financial prices move—cotton prices, wheat prices, railroad and Blue Chip stocks, dollar-yen exchange rates. I see a pattern in these price movements—not a pattern, to be sure, that will make anybody rich; I agree with the orthodox economists that stock prices are probably not predictable in any useful sense of the term. But the risk certainly does follow patterns that can be expressed mathematically and can be modeled on a computer. Thus, my research could help people avoid losing as much money as they do, through foolhardy underestimation of the risk of ruin. Thinking about markets as a scientific system, we may eventually craft a stronger financial industry and a better system of regulation.

A warning to readers here and now: Some of what I say has been embraced as economic orthodoxy in the past decade—but some of it remains contested, ridiculed, even vilified. When I publish in academic journals, as a scientist must, I often stir intense controversy. Each time, I have listened to the critics, rephrased my claims, gone back to my study to think and to my computers to analyze, and devised better, more-accurate models. Result: progress. Unavoidable side-effect: an element of complication. Indeed, I did not conceive of just one model of price variation, but several. Starting in 1963 and 1965 I devised two separate but incompatible models of behavior, succeeding at last in reconciling them in 1972. After a long detour through other fields of science, I resumed my financial research in 1997. This book guides the reader along the same winding journey of scientific discovery as I took. The goal: a better understanding of financial markets.

My oldest, best-corroborated insights now influence some of the mathematical models by which traders price options and banks evaluate risk. My scientific approach to markets has been emulated by a new generation of those who call themselves "econophysicists." And my latest models have been studied by a small but growing band of mathematicians, economists, and financiers in Zurich, Paris, London, Boston, and New York. I have no financial interest in their success or failure; I am a scientist, not a money man. But I wish them good fortune.

And I hope readers of this book, whether they agree or disagree with everything I say, will forsake, at least for a moment, the practical details of *why*. Instead, I hope they emerge from the book's pages with a greater fundamental understanding of *how* financial markets work, and of the great risk we run when we abandon our money to the winds of fortune.

The Study of Risk

There are many ways of handling risk. In the financial markets, the oldest is the simplest: "fundamental" analysis. If a stock is rising, seek the cause in a study of the company behind it, or of the industry and economy around it. Study harder, and predict the stock's next move. "Because" is the key word here: The price of a stock, bond, derivative, or currency moves "because" of some event or fact that more often than not comes from outside the market. World wheat prices rise because a heat wave desiccates Kansas or Ukraine. The dollar sinks because talk of war raises oil prices. This is all common sense. Financial newspapers thrive on it; they sell news and rank the importance of all the "becauses." Financial firms make an industry of it; they employ thousands of fundamental analysts, classified by genus into macroeconomic and sectoral, "top-down" and "bottom-up." Regulators codify and enforce it; they dictate what a company must tell its investors. The implicit assumption in all this: If one knows the cause, one can forecast the event and manage the risk.

Would it were so simple. In the real world, causes are usually obscure. Critical information is often unknown or unknowable, as when the Russian economy trembled in August 1998. It can be concealed or misrepresented, as during the Internet bubble or the Enron and Parmalat corporate scandals. And it can be misunderstood: The precise market mechanism that links news to price, cause to effect, is mysterious and seems inconsistent. Threat of war: Dollar falls. Threat of war: Dollar rises. Which of the two will actually happen? After the fact, it seems obvious; in hindsight, fundamental analysis can be reconstituted and is always brilliant. But before the fact, both outcomes may seem equally likely. So how can one base an investment strategy and a risk profile entirely on this one dubious principle: I can know more than anybody else?

In response, the financial industry has developed other tools. The second-oldest form of analysis, after fundamental, is "technical." This is a craft of recognizing patterns, real or spurious—of studying reams of price, volume, and indicator charts in search of clues to buy or sell. The language of the "chartists" is rich: head and shoulders, flags and pennants, triangles (symmetrical, ascending, or descending). The discipline, in disfavor during the 1980s, expanded in the 1990s as thousands of neophytes took to the Internet to trade stocks and insights. It truly thrives, however, in currency markets. There, all major "forex" houses employ technical analysts to find "support points," "trading ranges," and other patterns in the tick-by-tick data of the world's biggest and fastest market. And in the fun-house mirror logic of markets, the chartists can at times be correct. Sterling/dollar quotes really can approach a level advertised by the technical analysts, and then pull back as if hitting a solid wall—or accelerate as if bursting through a barrier. But this is a confidence trick: Everybody knows that everybody else knows about the support points, so they place their bets accordingly. It beggars belief that vast sums can change hands on the basis of such financial astrology. It may work at times, but it is not a foundation on which to build a global risk-management system.

And so was born what business schools now call "modern" finance. It emerged from the mathematics of chance and statistics. The fundamental concept: Prices are not predictable, but their fluctuations can be described by the mathematical laws of chance. Therefore, their risk is measurable, and manageable. This is now orthodoxy to which I subscribe—up to a point.

Work in this field began in 1900, when a youngish French mathematician, Louis Bachelier, had the temerity to study financial markets at a time "real" mathematicians did not touch money. In the very different world of the seventeenth century, Pascal and Fermat (he of the famous "last theorem" that took 350 years to be proved) invented probability theory to assist some gambling aristocrats. In 1900, Bachelier passed over fundamental analysis and charting. Instead, he set in motion the next big wave in the field of probability theory, by expanding it to cover French government bonds. His key model, often called the "random walk," sticks very closely indeed to Pascal and Fermat. It postulates prices will go up or down with equal probability, as a fair coin will turn heads or tails. If the coin tosses follow each other very quickly, all the hue and cry on a stock or commodity exchange is literally static—white noise of the sort you hear on a radio when tuned between stations. And how much the prices vary is measurable. Most changes, 68 percent, are small moves up or down, within one "standard deviation"—a simple mathematical yardstick for measuring the scatter of data—of the mean; 95 percent should be within two standard deviations; 98 percent should be within three. Finally—this will shortly prove to be very important—extremely few of the changes are very large. If you line all these price movements up on graph paper, the histograms form a bell shape: The numerous small changes cluster in the center of the bell, the rare big changes at the edges.

The bell shape is, for mathematicians, *terra cognita*, so much so that it came to be called "normal"—implying that other shapes are "anomalous." It is the well-trodden field of probability distributions that came to be named after the great German mathematician Carl Friedrich Gauss. An analogy: The average height of the U.S. adult male population is about 70 inches, with a standard deviation around two inches. That means 68 percent of all American men are between 68 and 72 inches tall; 95 percent between 66 and 74 inches; 98 percent between 64 and 76 inches. The mathematics of the bell curve do not entirely exclude the possibility of a 12-foot giant or even someone of negative height, if you can imagine such monsters. But the probability of either is so minute that you would never expect to see one in real life. The bell curve is the pattern ascribed to such seemingly disparate variables as the height of Army cadets, IQ test scores, or—to return to Bachelier's simplest model—the returns from betting on a series of coin tosses. To be sure, at any particular time or place extraordinary patterns can result: One can have long streaks of tossing only "heads," or meet a squad of exceptionally tall or dim soldiers. But averaging over the long run, one expects to find the mean: average height, moderate intelligence, neither profit nor loss. This is not to say fundamentals are unimportant; bad nutrition can skew Army cadets towards shortness, and inflation can push bond prices down. But as we cannot predict such external influences very well, the only reliable crystal ball is a probabilistic one.

Genius, in any time or clime, is often unrecognized. Bachelier's doctoral dissertation was largely ignored by his contemporaries. But his work was translated into English and republished in 1964, and thence was developed into a great edifice of modern economics and finance (and five Nobel Memorial Medals in economic science). A broader variant of Bachelier's thinking often goes by the title one of my doctoral students, Eugene F. Fama of the University of Chicago, gave it: the Efficient Market Hypothesis. The hypothesis holds that in an ideal market, all relevant information is already priced into a security today. One illustrative possibility is that yesterday's change does not influence today's, nor today's, tomorrow's; each price change is "independent" from the last.

With such theories, economists developed a very elaborate toolkit for analyzing markets, measuring the "variance" and "betas" of different securities and classifying investment portfolios by their probability of risk. According to the theory, a fund manager can build an "efficient" portfolio to target a specific return, with a desired level of risk. It is the financial equivalent of alchemy. Want to earn more without risking too much more? Use the modern finance toolkit to alter the mix of volatile and stable stocks, or to change the ratio of stocks, bonds, and cash. Want to reward employees more without paying more? Use the toolkit to devise an employee stock-option program, with a tunable probability that the option grants will be "in the money." Indeed, the Internet bubble, fueled in part by lavish executive stock options, may not have happened without Bachelier and his heirs.

Alas, the theory is elegant but flawed, as anyone who lived through the booms and busts of the 1990s can now see. The old financial orthodoxy was founded on two critical assumptions in Bachelier's key model: Price changes are statistically independent, and they are normally distributed. The facts, as I vehemently argued in the 1960s and many economists now acknowledge, show otherwise.

First, price changes are not independent of each other. Research over the past few decades, by me and then by others, shows that many financial price series have a "memory," of sorts. Today does, in fact, influence tomorrow. If prices take a big leap up or down now, there is a measurably greater likelihood that they will move just as violently the next day. It is not a well-behaved, predictable pattern of the kind economists prefer—not, say, the periodic up-and-down procession from boom to bust with which textbooks trace the standard business cycle. Examples of such simple patterns, periodic correlations between prices past and present, have long been observed in markets—in, say, the seasonal fluctuations of wheat futures prices as the harvest matures, or the daily and weekly trends of foreign exchange volume as the trading day moves across the globe.

My heresy is a different, fractal kind of statistical relationship, a "long memory." This is a delicate point to which a full chapter will be devoted later. For the moment, think about it by observing that different kinds of price series exhibit different degrees of memory. Some exhibit strong memory. Others have weak memory. Why this should be is not certain; but one can speculate. What a company does today—a merger, a spin-off, a critical product launch—shapes what the company will look like a decade hence; in the same way, its stock-price movements today will influence movements tomorrow. Others suggest that the market may take a long time to absorb and fully price information. When confronted by bad news, some quick-triggered investors react immediately while others, with different financial goals and longer time-horizons, may not react for another month or year. Whatever the explanation, we can confirm the phenomenon exists—and it contradicts the random-walk model.

Second, contrary to orthodoxy, price changes are very far from following the bell curve. If they did, you should be able to run any market's price records through a computer, analyze the changes, and watch them fall into the approximate "normality" assumed by Bachelier's random walk. They should cluster about the mean, or average, of no change. In fact, the bell curve fits reality very poorly. From 1916 to 2003, the daily index movements of the Dow Jones Industrial Average do not spread out on graph paper like a simple bell curve. The far edges flare too high: too many big changes. Theory suggests that over that time, there should be fifty-eight days when the Dow moved more than 3.4 percent; in fact, there were 1,001. Theory predicts six days of index swings beyond 4.5 percent; in fact, there were 366. And index swings of more than 7 percent should come once every 300,000 years; in fact, the twentieth century saw forty-eight such days. Truly, a calamitous era that insists on flaunting all predictions. Or, perhaps, our assumptions are wrong.

The Power of Power Laws

Examine price records more closely, and you typically find a different kind of distribution than the bell curve: The tails do not become imperceptible but follow a "power law." These are common in nature. The area of a square plot of land grows by the power of two with its side. If the side doubles, the area quadruples; if the side triples, the area rises nine-fold. Another example: Gravity weakens by the inverse power of two with distance. If a spaceship doubles its distance from Earth, the gravitational pull on it falls to a fourth its original value. In economics, one classic power law was discovered by Italian economist Vilfredo Pareto a century ago. It describes the distribution of income in the upper reaches of society. That power law concentrates much more of a society's wealth among the very few; a bell curve would be more equitable, scattering incomes more evenly around an average. Now we reach one of my main findings. A power law also applies to positive or negative price movements of many financial instruments. It leaves room for many more big price swings than would the bell curve. And it fits the data for many price series. I provided the first evidence in a 1962 research report, summarized by a brief published paper. The report showed that in the distribution of cotton price movements over the past century, the tails followed a power law; there were far too many big price swings to fit a bell curve. The same report continued with wheat prices, many interest rates, and railroad stocks—in other words, all the data I could locate in dusty library corners. Since then, a similar pattern has been found in many other financial instruments.

Economics is faddish. As in many scientific fields, so in the dismal science a consensus emerges about what is right and what is wrong, what research is worthy a doctoral thesis and what is not. I have run counter-trend most of my professional career. In the 1960s, most theoretical economists were lionizing Bachelier and his heirs. The next decade, Wall Street embraced their theories. They were the intellectual foundation for stock-index funds, options exchanges, executive stock options, corporate capital-budgeting, bank risk-analysis, and much of the world financial industry as we know it today. Throughout this time, I was being heard, but as a near-lone voice denouncing the flaws in the logic. By the late 1980s and 1990s, however, I was no longer alone in seeing those flaws. The financial dislocations convinced many professional financiers that something was wrong. Warren E. Buffett, the famously successful investor and industrialist, jested that he would like to fund university chairs in the Efficient Market Hypothesis, so that the professors would train even more misguided financiers whose money he could win. He called the orthodox theory "foolish" and plain wrong. Yet none of its proponents "has ever said he was wrong, no matter how many thousands of students he sent forth misinstructed. Apparently, a reluctance to recant, and thereby to demystify the priesthood, is not limited to theologians."

However dogmatic the professors, the practical men of Wall Street did eventually open to new ideas. My principal objections—that prices do not follow the bell curve and are not independent—were heeded, and hundreds of economists and market analysts have by now documented their validity. But despite recognition of the problem, the old methods have surprising staying-power. The "classical" formulae of Bachelier and his heirs—how to build an investment portfolio, to evaluate the financial value of a new factory, to judge the riskiness of a stock—remain on the curriculum at hundreds of business schools around the world and are a standard part of the Chartered Financial Analyst exams administered to thousands of young brokers and bankers. They remain part of the orthodoxy of Wall Street professionals, too. For instance, the "Black-Scholes" formula for valuing a Merrill or GM executive's stock options was long the gold standard; only in 2004 did U.S. regulators officially countenance other formulae. Why such reluctance to change? The old methods are easy and convenient. They work fine, it is argued, for most market conditions. It is only in the infrequent moments of high turbulence that the theory founders—and at such moments, who can guard against a hostile takeover, a bankruptcy or other financial act of God? Such reasoning, of course, is little comfort to those wiped out on one of those "improbable," violent trading days.

But the financial industry is supremely pragmatic. While it may genuflect to the old icons, it invests its research dollars in the search for newer, better gods. "Exotic" options, "guaranteed-return" products, "value-at-risk" analysis, and other Wall Street creations have all benefited from this search. Central bankers, too, are pragmatic. After years of accepting the old ways, they have been pushing since 1998 for new, more realistic mathematical models by which a bank should evaluate its risk. These so-called Basle II rules will force many banks to change the way they calculate how much capital they set aside as a cushion against financial catastrophe. In response, economists have been rushing to oblige with new ideas and new models. Many, with such unattractive names as GARCH and FIGARCH, just patch the old models. Others start from scratch, rejecting all the old assumptions. Behavioral economists study mar-

kets as B. F. Skinner studied humans: as organisms that input information and output behavior according to rules to be deduced. In
this spirit, some researchers have wired professional traders to
measure skin resistance, EEG patterns, and pulse rates, in search of
the biological imperative behind a "buy" order. And there is com-
puter-intensive finance. Wall Street has long been the computer
industry's biggest customer, unleashing "genetic algorithms," "neu-
ral networks," and other computational techniques on the market
in hopes that silicon intelligence can find profitable patterns where
carbon-based life forms cannot.

This "post-modern" finance has yet to yield real success. Nobody
has hit the jackpot.

A Game of Chance

So, as Lenin's revolutionary manifesto put it: What is to be done?

As preparation, play a game.

On the facing page you see four price charts of the kind you
would find in a brokerage-house report, but with the identifying
dates and values removed. Two of the charts are real chronicles of
the price of a real financial instrument—name also removed. Two
are forgeries, entirely fictitious series of numbers, generated using
different theoretical models of how markets work. Ignore whether
they trend up or down. Focus on how they vary from one moment
to the next. Which are real? Which fake? What rules were used to
draw the fake?

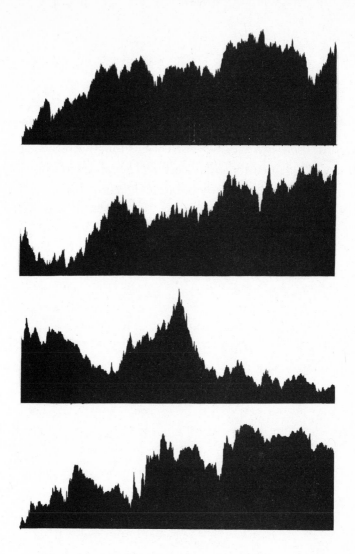

Four charts: Which are real, which are fake?

All fairly similar, many readers will say. Indeed, stripped of legends, axis labels, and other clues to context, most price "fever charts," as they are called in the financial press, look much the same. But pictures can deceive better than words.

For the truth, look at the next set of charts. These show, rather than the prices themselves, the changes in price from moment to moment. Now, a pattern emerges, and the eye is smarter than we normally give it credit for—especially at perceiving how things change.

The worst fake stands out from the rest, like a criminal in a police line-up. It is the second chart, which shows prices varying more or less uniformly over time. It was generated by the orthodox random-walk model. The size of most price changes varies within a narrow range, corresponding to the central portion of the bell curve mentioned earlier. True, the chart also shows bigger fluctuations, or outliers—but they barely stand up from the bulk of changes, as taller strands of grass rise above the average height of an unmown lawn.

Compare this fake chart with the two real ones, numbers 1 and 3. The top-most charts the relative price changes of IBM stock from 1959 to 1996; the third one charts the relative changes in the dollar/Deutschemark exchange rate. In these and all other real charts, price swings are highly erratic. The large ones are numerous and cluster together. Here, the appropriate analogy is no longer to grass, but to a forest of trees of all sizes—some gigantic. Another analogy is to the distribution of stars. They are not uniformly distributed throughout the universe. Instead they cluster into galaxies, then into galaxy clusters, in a hierarchy both random and ordered. Mathematically speaking, much the same thing is going on in these stock-price charts.

That leaves Chart No. 4—the ringer in this game. It is a fictitious series of price changes generated using my latest model of how financial markets work. It faithfully simulates the "volatile volatility" of the real charts—and, whether in financial modeling or weather forecasting, the proof of any model lies in its results. In times past, the predictions of models were expressed in a few numbers or diagrams. I pioneered the use of the computer to express the predictions of my models in this unique graphical form, a kind of forgery of reality. Here, the underlying model is called fractional Brownian motion in multifractal time. Though the name is forbid-

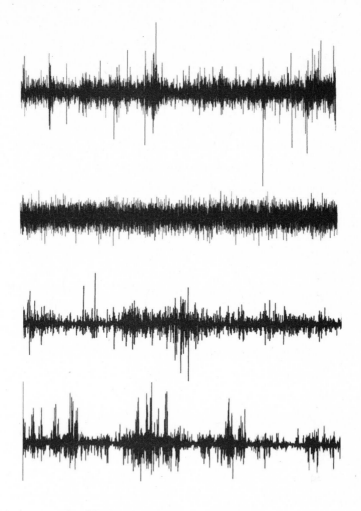

**The "daily changes" in the four charts.
Again, which are fake?**

ding, later chapters will elaborate and show the model to be extremely parsimonious.

How does it work? It is based on my fractal mathematics, which subsequent chapters will elucidate. It is a model still in development. What I know cannot yet be used to pick stocks, trade derivatives, or value options; time, and further research by others, will determine whether it ever can. But to be able to imitate reality is a form of understanding, and as such, the multifractal model already offers some immediate insights into how markets work. Like the popular-finance press, I can boil some of them down to five "rules" of market behavior—concepts that, if grasped and acted upon, can help lessen our financial vulnerability.

Rule I. Markets are risky.

Extreme price swings are the norm in financial markets—not aberrations that can be ignored. Price movements do not follow the well-mannered bell curve assumed by modern finance; they follow a more violent curve that makes an investor's ride much bumpier. A sound trading strategy or portfolio metric would build this cold, hard fact into its foundations. Exactly how depends on the resources, talents, and stomach for risk of the individual; as ever, differing opinions make a market. But already, the mere knowledge that markets vary wildly is useful. It can be—and increasingly is—used in computer simulations to "stress-test" a portfolio, to play a wider and darker range of "what-if?" games on paper, before committing hard cash to a trading strategy. Thus, a cautious investor can build a portfolio with greater security than the standard models suggest. An aggressive trader can be better prepared to pounce on moments of high volatility. And a prudent market regulator can be more alert to urgent problems—thereby averting financial catastrophe and macroeconomic harm. Some commentators have called for a "Richter scale" of market turbulence; like that famous measure of earthquake intensity, its financial analog would rank market tremors and provide a scale for regulators to judge the severity of impending problems. Forewarned is forearmed.

Rule II. Trouble runs in streaks.

Market turbulence tends to cluster. This is no surprise to an experienced trader. In financial dealing-rooms across the world, the first fifteen minutes of trading each morning are critically important; it is when experienced traders, staring at their screens, take the temperature of the market. They know that when a market opens choppily, it may well continue that way. They know that a wild Tuesday may well be followed by a wilder Wednesday. And they also know that it is in those wildest moments—the rare but recurring crises of the financial world—that the biggest fortunes of Wall Street are made and lost. They need no economists to tell them all this. But their intuition, not included in the standard model of efficient markets, is entirely validated by the multifractal model.

Rule III. Markets have a personality.

Prices are not driven solely by real-world events, news, and people. When investors, speculators, industrialists, and bankers come together in a real marketplace, a special, new kind of dynamic emerges—greater than, and different from, the sum of the parts. To use the economists' terms: In substantial part, prices are determined by *endogenous* effects peculiar to the inner workings of the markets themselves, rather than solely by the *exogenous* action of outside events. Moreover, this internal market mechanism is remarkably durable. Wars start, peace returns, economies expand, firms fail— all these come and go, affecting prices. But the fundamental process by which prices react to news does not change. A mathematician would say market processes are "stationary." This contradicts some would-be reformers of the random-walk model who explain the way volatility clusters by asserting that the market is in some way changing, that volatility varies because the pricing mechanism varies. Wrong. A striking example: My analysis of cotton prices over the past century shows the same broad pattern of price variability at the turn of the last century when prices were unregulated, as there was in the 1930s when prices were regulated as part of the New Deal.

Rule IV. Markets mislead.

Patterns are the fool's gold of financial markets. The power of chance suffices to create spurious patterns and pseudo-cycles that, for all the world, appear predictable and bankable. But a financial market is especially prone to such statistical mirages. My mathematical models can generate charts that—purely by the operation of random processes—appear to trend and cycle. They would fool any professional "chartist." Likewise, bubbles and crashes are inherent to markets. They are the inevitable consequence of the human need to find patterns in the patternless.

Rule V. Market time is relative.

There is what one may call a relativity of time in financial markets. Early on, but mostly when developing the multifractal model, I came to think of markets as operating on their own "trading time"—quite distinct from the linear "clock time" in which we normally think. This trading time speeds up the clock in periods of high volatility, and slows it down in periods of stability. Mathematically, I can write an equation showing how one time frame relates to the other and use it to generate the same kind of jagged price series that we observe in real life. This is how the successful forgery shown among the previous charts was made. It is almost as if dealing rooms need, besides the standard row of wall-clocks showing the time in Tokyo, London, and New York, a fourth clock showing "Greenwich Market Time."

This last point highlights an important subtext of this book: Market professionals know far more than they even realize. Professional traders often speak of a "fast" market or a "slow" one, depending on how they judge the volatility at that moment. They would quickly recognize, and affirm, the concept of trading time. Likewise, a bit of market folk-wisdom holds that all charts look alike: Without the identifying legends, one cannot tell if a price chart covers eighteen minutes, eighteen months, or eighteen years. This will be expressed by saying that markets scale. Even the financial press scales: There are annual reviews, quarterly bulletins, monthly newsletters, weekly magazines, daily newspapers, and tick-by-tick electronic newswires and Internet services. Market folklore and anecdote, of course, cannot confirm the multifractal model; only rigorous statistical analysis can do that. But the folklore does signal that the model is on the right track.

The multifractal model also has many implications for practical finance. As indicated, portfolio theory needs rethinking; options need revaluing; trading strategies need review. A small example: "stop-loss" orders are imperfect, to put it mildly. Many investors or traders leave instructions to close a position when a price hits a particular target. But as many have learned to their grief, when prices are really flying, they typically whiz past the target so fast that even the most attentive broker cannot execute the "sell" orders fast enough. Result: Greater losses, or smaller profit, than the investor intended. Another example: the mathematics of this model offers some potentially new yardsticks to measure volatility and risk. Instead of the standard deviations and "betas" of conventional finance, one can imagine new scales based on two new variables to be described later in this book: the H exponent of price dependence, and the α parameter characterizing volatility. A few fund managers have experimented with these concepts. They often call it chaos theory—though strictly speaking, that is marketing language riding on the coattails of a popular scientific trend. In reality, the mathematics is still young, the research barely begun, and reliable applications still distant.

So *caveat emptor*: This book will not make you rich. Bookseller: Do not put it on the same shelf with the "How to Make a Million in the Market" volumes. If it fits any genre, it is that of popular science. It explains a new, and important, way of looking at the world—in this case, the financial world. It attempts to do so using common English, with as few formulae and as little mathematical jargon as possible—or at least, with no jargon unexplained. That is because I aim to stimulate broader debate about financial-market modeling. It is a debate that has, hitherto, been confined to the rarefied circles of economics-minded mathematicians, or of mathematically inclined economists. The underlying mathematics is, frankly, forbidding—the primary reason why, when I first began publishing in the 1960s and 1970s, few mainstream economists were inclined to listen. But the extraordinary tumult and noise of this fin de siècle market turmoil are opening the ears of many who previously affected deafness.

Research in this field has far to go. It took more than sixty years after Bachelier's thesis for economists to formulate properly the Efficient Market Hypothesis, and another decade beyond that for their work to find valuable applications in the real world of zero-coupons and call options. With fractals, we are only a few short decades from the origin. But they already illumine some profound truths of finance and economics. Chief among these is the paramount importance of risk.

We have been mis-measuring risk.

Greater knowledge of a danger permits greater safety. For centuries, shipbuilders have put care into the design of their hulls and sails. They know that, in most cases, the sea is moderate. But they also know that typhoons arise and hurricanes happen. They design not just for the 95 percent of sailing days when the weather is clement, but also for the other 5 percent, when storms blow and their skill is tested. The financiers and investors of the world are, at the moment, like mariners who heed no weather warnings. This book is such a warning.

CHAPTER II

By the Toss of a Coin or the Flight of an Arrow?

FOR MOST PEOPLE, chance is a familiar but unexamined idea, a word with many separate meanings. They speak of the chance of winning the lottery, or the chance of being in a plane crash; they mean a simple number, the odds of something happening. Or they speak of a chance encounter, by which they mean unplanned, unanticipated. When they are investing, they have yet another meaning. They speak of the chance of losing money; here, chance is a menace, a risk. It is the thing that upsets their investment plans, makes them poor where they hoped to be rich. They try to weigh risks, comparing stocks with bonds, real estate with Treasuries. Most people have no idea how to do that systematically and numerically, but they accept that chance is, somehow, involved in their personal investments. Considering the alternative—that they have only themselves to blame for a lousy investment—bad luck makes a handy scapegoat.

But can chance describe not just their personal misfortunes, but the operations of the market overall? Bunk, say some. We live in the real world of brokers, investors, and hard cash, not abstract proba-bility. IBM stock rose by $1 a share because the company announced it signed more computer-service contracts than expected, and so 5,218 real people, some calculating and some impulsive, some greedy and some prudent, ordered 12,542,300 real IBM shares with $768,016,733 in real cash. It is cause-and-effect, the very model of determinism. No luck about it, whatsoever. Sure, it is difficult to reconstruct who did what and why to make the price rise, and harder still to forecast whether it will keep rising; that is what bro-kers are for. But it is nonsense to suggest that IBM stock rose by chance. Dice fall by chance. Roulette wheels spin by chance. But IBM shares, the euro-dollar exchange rate, and wheat prices do not rise or fall by the mathematical rules of chance.

Indeed, they do not—but they can be described as if they do. And that subtle distinction, of thinking about prices *as if* they were gov-erned by chance, has been the dominant, fructifying notion of finan-cial theory for the past one hundred years. On its foundation was built the modern, global financial industry. Portfolio management, trading strategy, corporate finance—all have been shaped by the chain of assumptions and deductions that succeeding generations of economists and mathematicians have forged from this paradoxical notion of chance.

I am, of course, a true believer in the power of probability. I have seen it and applied it in economics, physics, information theory, metallurgy, meteorology, neurology, anatomy, taxonomy, and many other seemingly improbable fields. As a graduate student at the University of Paris more than fifty years ago, I wrote my doctoral thesis on an ignored byway of applied probability: the power law that rules the mathematical frequency with which individual words occur in common language. With such a background I would hardly be one to refute the usefulness of probability theory in yet another field, finance. In financial markets, God can *appear*, any-way, to play with dice. What I know is that the ruler of chance can create what I call several distinct "states" or types of chance. And what I contest is the way today's financial theorists, in their class-rooms and their writings, calculate the odds. It may seem to some an academic quibble—but as will be seen, it can be the difference between winning and losing a fortune.

To grasp this crucial point—indeed, the spine of this whole book—it helps to go back to basics. This chapter starts with a look at two sharply different probabilistic tools. The next chapter tells the story of how modern financial theory was built. Then that con-struction is examined critically. Finally, I propose a plan for repairs. As will be seen, I am not a Luther fomenting schism in the Church. I am an Erasmus who, through study, reason, and good humor, tries to talk some sense. My aim: To change the way people think, so that reform may go forward.

Chance in Finance

Why even talk about chance in financial markets? The very idea clashes with every intuition we have about the way society, com-merce, and finance work. In reply, consider two ways of looking at the world: as a Garden of Eden or as a black box.

The first is cause-and-effect, or deterministic. Here, every par-ticle, leaf, and creature is in its appointed place, and, if only we had the vast knowledge of God, everything could be understood and predicted. Scientists once thought this way. Two centuries ago, when new telescopes and new math were opening the mod-ern study of astronomy, the great French mathematician, the Marquis Pierre-Simon de Laplace, asserted that he could predict the future of the cosmos—if only he knew the present position and velocity of every particle in it. This view, carried over into mar-kets, would be a full-employment act for the world's financial ana-lysts and economists. They could tell you whether inflation would rise, whether interest rates would fall, and which stocks to buy and sell—if only they had enough good data, if only they had good enough computers, if only there were enough of them earning good salaries.

Enough. How realistic is that? We cannot know everything. Physicists abandoned that pipedream during the twentieth century after quantum theory and, in a different way, after chaos theory. Instead, they learned to think of the world in the second way, as a black box. We can see what goes into the box and what comes out of it, but not what happens inside; we can only draw inferences about the odds of input A producing output Z. Seeing nature through the lens of probability theory is what mathematicians call the stochastic view. The word comes from the Greek *stochastes*, a diviner, which in turn comes from *stokhos*, a pointed stake used as a target by archers. We cannot follow the path of every molecule in a gas; but we can work out its average energy and probable behavior, and thereby design a very useful pipeline to transport natural gas across a continent to fuel a city of millions.

If the physical world is so uncertain, so difficult to know precisely, then how much more uncertain and unknowable must be the world of money? Finance is a black box covered by a veil. Not only are the inner workings hidden, but the inputs are also obscured, by bad economic data, conflicting news reports, or outright deception. What coefficient of correction should I apply to a broker's self-serving stock tip? And then there is the most confounding factor of all, anticipation. A stock price rises not because of good news from the company, but because the brightening outlook for the stock means investors anticipate it will rise further, and so they buy. Anticipation is a feature unique to economics. It is psychology, individual and mass—even harder to fathom than the paradoxes of quantum mechanics. Anticipation is the stuff of dreams and vapors.

Yet in economics, there must be scores of academic journals in which scholars struggle to follow Laplace, trying to model the inner workings of the economy in all its splendid detail. They work from vast databases of prices and production. They make assumptions about human behavior, and so hypothesize intricate relations among the rate of savings, the rate of interest, and other economic variables. They try to seize in a moment a very complicated thing.

A contrary approach, macroscopic instead of microscopic, stochastic instead of deterministic, would be more fruitful. The theory of magnets is worth mentioning here. When temperature rises above a certain critical level called the Curie point, magnetism disappears. As the metal is cooled back down below that point, magnetism returns. This, in a matter of nanoseconds. How? Despite two centuries of research, we still do not know precisely—but we have macroscopic theories for it that work very well. In flat magnets a chemist who was also a mathematician and physicist, Lars Onsager, drew immense insights from a ludicrously simple model. Imagine a magnet's sub-atomic particles as arrayed in a grid like traffic lights on the street corners of New York City. Each light can be in one of two states, called "up" or "down" spin. When they are more or less aligned, you get more or less strong magnetism; when they are all working at cross purposes, you lose it. As the temperature rises, extra energy swamps the grid and knocks the spins out of alignment. As it falls, neighboring lights start cooperating with one another again and try to get back into synch. The math for it is straightforward in principle, but in practice, devilish enough for a Nobel Prize. Now, this is an overly simple theory—simpleton, in fact. Fortunately, how and why each individual particle interacts with the next happens to matter less than one may think. We can use this theory to design electrical generators, computer disks, and thousands of other very practical devices.

Still, the idea of chance in markets is difficult to grasp, perhaps because, unlike the anonymous particles in a magnet or molecules in a gas, the millions of people who buy and sell securities are real individuals, complex and familiar. But to say the record of their transactions, the price chart, can be described by random processes is not to say the chart is irrational or haphazard; rather, it is to say it is unpredictable. Again, word derivations are helpful. The English phrase "at random" adapts a medieval French phrase, *à randon*. It denoted a horse moving headlong, with a wild motion that the rider could neither predict nor control. Another example: In Basque, "chance" is translated as *zoria*, a derivative of *zhar*, or bird. The flight of a bird, like the whims of a horse, cannot be predicted or controlled.

We can think of financial prices in much the same way: not predictable, not controllable. Under such circumstances, the best we can do is evaluate the odds for or against some outcome: a stock rising a certain amount this year, an option coming into the money, or an exchange rate holding steady through the next corporate budget cycle. To use the tools of probability is not to say chance governs global commerce and finance. Sure, after the fact, with enough time and effort, we can piece together a tolerable cause-and-effect story of why a price moved the way it did. But who cares? It is too late by then. Fortunes have been gained and lost. Before the fact, in the real world of fast markets, veiled motives, and uncertain outcomes, probability is the only tool at our disposal.

Chance, Simple or Complex

But how, you may ask, can the tools of probability describe the amazing richness of a stock chart?

First and foremost, random need not mean simple. There is more to probability than coins and dice. In the hands of a mathematician, even the most trivial random process—for example, a coin game—can generate surprising complexity, baroque detail, and highly structured behavior. One of the founders of modern probability theory, the late Russian mathematician Andrei Nikolaievitch Kolmogorov, wrote, "the epistemological value of probability theory is based on the fact that chance phenomena, considered collectively and on a grand scale, create a non-random regularity." Sometimes this regularity can be direct and awesome, at other times strange and wild.

For example, consider the old game of tossing a coin. It has been popular among theoreticians since the days of the Bernoulli brothers, a prolific family of eighteenth-century mathematicians from Basel whose studies helped found the field of probability. Imagine that Harry wins a Swiss franc on heads, and his brother Tom wins one on tails. (Past mathematicians called them Peter and Paul. But I could never remember which was which.) Each toss is pure luck. But after these three centuries of playing the game, millions and millions of times, each brother has every reason to expect to have won half of the time. Such is the dictate of the law of large numbers, a common-sense notion also approved by mathematicians: If you repeat a random experiment often enough, the average of the outcomes will converge towards an expected value. With a coin, heads and tails have equal odds. With a die, the side with one spot will come up about a sixth of the time. This is what Kolmogorov meant.

But other aspects of the game get more complicated. At any particular moment, one brother may have accumulated far more winnings than the other. Look at the full record of a coin-tossing experiment on the following page—10,000 simulated tosses. It is due to an eminent mathematician I knew well, Willy Feller, who in 1950 wrote a probability textbook widely used at one time. After each toss, he charted Harry's cumulative winnings or debts. An erratic, but pronounced, pattern appears: A few long, up-and-down cycles stand out, while many shorter cycles ride on top of them. The "zero-crossings"—the moments when the imaginary purses of Harry or Tom go back to the empty state at which they started—are not uniformly spread but cluster together. It is structure of an irregular kind.

All those years ago, when this diagram was first published, few readers heeded it. But I spent hours examining it, dreaming on it, trying to discern the chance patterns and processes behind it. At first glance, how much like a stock chart is this? "Chartists" spend their days studying financial graphs, spotting head-and-shoulder patterns, identifying compression periods or support levels, and then confidently advising their clients to buy or sell. Would they spot the difference if I slipped one of these coin-tossing charts into their folders? Should I expect a call from one, advising me to buy?

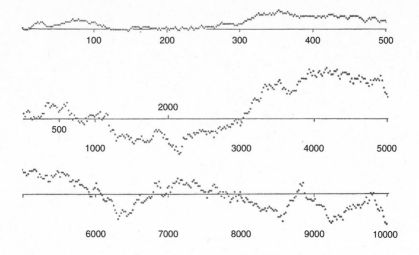

The record of 10,000 coin tosses. These charts, adapted from Feller 1950, show how far a coin-tosser's winnings can rise or fall from the expected average of zero (the horizontal lines). The top diagram shows the first 500 throws in detail. The lower two, placed end to end, cover 10,000 throws. The main point: A complex pattern can appear to emerge from even the simplest random process.

A key point in my work: Randomness has more than one "state," or form, and each, if allowed to play out on a financial market, would have a radically different effect on the way prices behave. One is the most familiar and manageable form of chance, which I call "mild." It is the randomness of a coin toss, the static of a badly tuned radio. Its classic mathematical expression is the bell curve, or "normal" probability distribution—so-called because it was long viewed as the norm in nature. Temperature, pressure, or other features of nature under study are assumed to vary only so much, and not an iota more, from the average value. At the opposite extreme is what I call "wild" randomness. This is far more irregular, more unpredictable. It is the variation of the Cornish coastline—savage promontories, craggy rocks, and unexpectedly calm bays. The fluctuation from one value to the next is limitless and frightening. In between the two extremes is a third state, which I call "slow" randomness.

Think about the three—mild, slow, and wild—as if the realm of chance were a world in its own right, with its own peculiar laws of physics. Mild randomness, then, is like the solid phase of matter: low energies, stable structures, well-defined volume. It stays where you put it. Wild randomness is like the gaseous phase of matter: high energies, no structure, no volume. No telling what it can do, where it will go. Slow randomness is intermediate between the others, the liquid state. I first proposed some of my views of chance in 1964 in Jerusalem, at an International Congress of Logic and Philosophy of Science. Since then, I have much expanded the theory and shown it to be critical to understanding financial markets in their proper light. As will be seen, the standard theories of finance assume the easier, mild form of randomness. Overwhelming evidence shows markets are far wilder, and scarier, than that.

The "Mild" Form of Chance

The most familiar type of randomness, expressed by the bell curve, first came into focus two centuries ago. From the start, its theory was both influential and controversial. Indeed, its discovery stirred a dispute over authorship—oft-told but worth repeating here—between an especially eminent mathematician, Adrien-Marie Legendre, and one of the greatest of all times, Carl Friedrich Gauss.

As the nineteenth century began, the calculation of celestial orbits was at the cutting edge of mathematical research. Improved telescopes were yielding new data on the heavens; and Newton's law of gravity provided the lens to interpret that data. But, as had been known as far back as Tycho Brahe in the late sixteenth century, telescope observations were prone to grievous error. There was the systematic error that arose from flaws in the instruments: an imperfectly ground lens, an uneven mounting. This kind of error could be explained, measured, and compensated for. Then there was the occasional error that could not be controlled: varying atmospheric conditions, tremors in the earth, or inebriated observatory assistants. This uncontrollable kind of error greatly complicated the task of calculating an orbit of a newly sighted comet or planet.

Like most great mathematicians until comparatively recent times, Legendre and Gauss had broad professional interests. Legendre in Paris rewrote Euclid's famous principles of geometry into what became a standard text in the field, wrote the first full-length treatise on number theory, and in the Napoleonic age helped precisely draw the map of France. Gauss in the north German Kingdom of Hanover (whose monarch had risen to the far richer throne in London) had been a child prodigy, a laborer's son who could count before he could speak and who developed his first famous mathematical proof, in geometry, when he was eighteen. Nearly every field he touched was the better for it: prime numbers, algebraic functions, infinite series, probability, topology. With a colleague, he designed the first electric telegraph. Like Legendre, he was a busy map surveyor. He calculated from meager data the orbits of several newly discovered planetoids. Indeed, his computational speed was legendary: The ten hours in which he determined and checked the orbit of one planetoid, Vesta, would have been, for a lesser man, several days of laborious calculation, tabular reference, and proofreading.

It was in astronomy that the two men clashed. In 1806, Legendre published a treatise on the calculation of orbits that included a supplement entitled, "On the method of least squares." It dealt with a common problem: how to find the "true" value of an orbit, or any other natural phenomenon, from a scattering of error-prone observations. The method was simple: Take a guess at the true value, and calculate how far away from it each observation is—the error. Then square each error and add them all together. Then take another guess at the true value, and see if the new squared errors are any smaller. Then do it again, and again. The "least-squares" estimate yields errors with the smallest sum of squares; it is the value that fits closest to all the observations. It was an effective method, immediately recognized as handy and, even today, used regularly in every form of physical research from astronomy to biology. But three years after Legendre, Gauss wrote about a similar method without acknowledging the Frenchman's work. Legendre protested. Gauss was always loath to waste time quarreling with other mathematicians. He did not respond directly but assured colleagues that he had thought of the method himself when he was eighteen years old and had used it repeatedly in his astronomical calculations. Laplace tried to mediate, to no avail. In the end, both men were credited with the discovery. Proof of Gauss's priority, found later in his voluminous notebooks, is somewhat controversial, but he clearly saw much deeper meaning in the method than did Legendre.

Let us return to the coin-toss game. Say Harry or Tom keeps a record—such as Feller's diagram reproduced earlier—of the deviations from the expected average of zero. Like in tennis, divide the game into "sets," each made of one million tosses, and record how

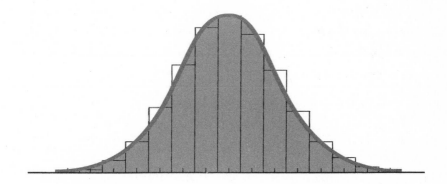

The bell curve. Harry is betting on heads coming up, and at each "set" of one million tosses keeps a record of his cumulative gain or loss. The height of the curve represents how often each type of outcome occurs. Most of the time, his winnings per set are small and get plotted into the fat center of this curve. Only occasionally, they are enormous— and appear on the skinny positive and negative "tails" of the curve. This is the distribution of a random process often called "normal."

much Harry won during the first set, the second, and so on. The size of the per-set purse will vary greatly, of course. It will often be about zero. But often, theory suggests, it will range in the favor of one brother or another—"typically," by 1,000 tosses. And on rare occasion, the "error," or deviation from the average they expect the coin to produce, will be far, far greater. If the brothers then graph the results in a "histogram" with a different-height bar for the number of times each score occurred, then the bars will start to form a familiar pattern. The numerous small winnings group around the expected average, zero—the tall center of the chart. The rare, fat purses go to the two extreme edges. Trace across the tops of all the bars, and you see the profile of the bell curve emerging.

If you study that bell curve, as did Gauss, some surprising facts arise. First, assume several games are going at once. While Harry and Tom play with the coin, their cousins are throwing dice and their friends are dealing cards. The players in each game expect a different average outcome; but for each, the graph of how their winnings per set differ from that average has the same general bell shape. Some bells may be squatter, and some narrower. But each has the same mathematical formula to describe it, and requires just two numbers to differentiate it from any other: the mean, or average, error, and the variance or standard deviation, an arbitrary yardstick that expresses how widely the bell spreads.

Now, this is all very convenient, in fact, simpler than most situations that occur in physics. One formula that includes two numbers as parameters can describe a vast range of human experience. Indeed, the common IQ test is deliberately designed to produce a bell curve of scores. The average IQ is, by definition, 100 points, the center of the bell. Then, 68 percent of the population has an IQ within one ten-point standard deviation of the mean, or between 90 and 110 points. About 95 percent are within two standard deviations of the mean, between 80 and 120 points. And 98 percent are within three standard deviations, also called sigma for the Greek letter, σ, used to write it. As sigma grows, the odds of being inside the bell rapidly approach 100 percent, while the odds of being out-side—an "outlier"—approach zero; an equation can estimate those odds. But that is not all. If you charted the IQ of every other person in a country rather than the entire population, you would still get a bell curve. If the verbal and mathematical test scores are independent and each follows a bell curve, so does the sum of the scores. Of course, the combined average score and its spread would have changed, but the basic properties of the curve would be the same.

In short, the normal curve is indestructible. It is mathematical alchemy. It is what you inevitably get if you combine lots of little variations, each one independent from the last, and each one negligible when compared to the total. No one individual matters much to the total IQ curve; no one coin toss matters much to Harry and Tom's game. But cumulatively, over time or across a population, the way the results vary forms a regular and predictable pattern. The data points are grains of sand on a shoreline, blades of grass in a lawn, electrons moving along a copper wire.

The Blindfolded Archer's Score

Now, this is a convenient way to look at the world, but is it the only way? Not at all. Late in his long life, the nineteenth-century French mathematician Augustin-Louis Cauchy thought of an especially tricky one. It was, when I was younger, viewed as interesting—but unrealistic and contrived. My work made it very real.

I think the theory best imagined in terms of an archer standing before a target painted on an infinitely long wall. He is blind-folded and consequently shoots at random, in any direction. Most of the time, of course, he misses. In fact, half of the time he shoots away from the wall, but let us not even record those cases. Now, had his recorded misses followed the mild pattern of a bell curve, most would be fairly close to the mark, and very few would be very wide of it. Suppose he shot arrows long enough, in successive "sets." For each set, he could calculate an average error and standard deviation—even give himself a score for blindfolded archery. But our archer is not in the land of the bell curve; his misses are not mild. All too often, his aim is so bad that the arrow flies almost parallel to the wall and strikes hundreds of yards from the target, or even a mile, if his arm is strong enough. Now, after each shot, let him try to work out his average target score. In the Gaussian environment, even the wildest shots have a negligible contribution to the average. After a certain number of strikes, the running average score will have settled down to one stable value, and there is practically no chance the next shot will change that average perceptibly. But the Cauchy case is completely different. The largest shot will be nearly as large as the sum of all the others. One miss by a mile completely swamps 100 shots within a few yards of the target. His scores for blindfolded archery never settle down to a nice, predictable average and a consistent variation around that average. In the language of probability, his errors do not converge to a mean. They have infinite expectation, hence also infinite variance.

Cauchy's is a totally different way of thinking of the world than Gauss's. The errors are not distributed as near-uniform grains of sand; they are a composite of grains, pebbles, boulders, and mountains. The practical importance of the distinction first became recognized through my work, but its existence was noted long ago. In 1853, the weekly printed proceedings of the French Academy of Sciences record a debate on the subject between Cauchy and another mathematician, Irénée Bienaymé. In effect, Cauchy observed that our archer's score challenged what was already by his time a casual, unreflective use of Gauss's formulae for nearly every measurement problem in science. Bienaymé retorted that the method was not merely convenient, but also reflected a fundamental truth about probability. Cauchy's fanciful error formula, he argued, was an unnatural oddity; if it ever occurred, a scientist would spot it immediately:

The observations themselves would warn the least attentive observer. As the large-value errors would have to have a noticeably large probability, from the start they would present themselves, if not as often as the others, then at least in as large a proportion. Thus, you would have frighteningly discordant observations. And there is no doubt that they would be rejected and that the instruments, or the observation process, would be submitted to profound correction. . . . An instrument governed by such a [Cauchy] law of probability would never be put on sale by an ordinary craftsman. One could not even imagine a firm that would manufacture one.

Comptes Rendus de l'Académie des Sciences, Aug. 29, 1853

Such has been the argument of most mathematicians and scientists ever since: Gaussian math is easy and fits most forms of reality, or so it seems. But with the sharp hindsight provided by fractal geometry, the Gaussian case begins to look not so "normal," after all. It was so-called only because science tackled it first; as ever in science, there is a healthy opportunism to begin with the problems easiest to handle. But the difference between the extremes of Gauss and of Cauchy could not be greater. They amount to two different ways of seeing the world: one in which big changes are the result of many small ones, or another in which major events loom disproportionately large. "Mild" and "wild" chance, described earlier, are my generalizations from Gauss and Cauchy.

You can see analogs of this dichotomy all around. In history, modernists argue that the course of human events is shaped by many trends, economic and social, enacted in the lives of millions of forgotten individuals; the historian's task is to trace these trends. By contrast, traditionalists, now coming back into fashion, contend that history was shaped and dominated by a few great men, Caesar or Napoleon, Newton or Einstein, for example. In the first, mild view, the birth or death of no single individual is crucial to the story of mankind; in the second, wild view, it most certainly is. Another

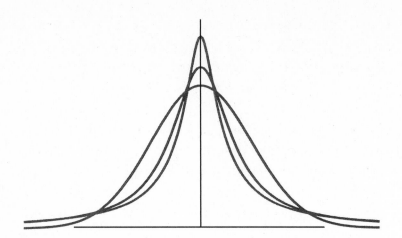

Three diagrams. In this figure, the famous bell curve is combined with two others of clearly different properties. The flattest curve is the bell, and its tails are so short that the horizontal axis had to be cut to avoid hiding them. The most peaked curve has extremely long tails. This is the Cauchy curve, that gives the distribution of the scores of our blindfolded archer. The intermediate curve will serve later in this book to represent the distribution of price increments of cotton.

example: Under a microscope, the edge of a sharp razor blade looks a bit ragged. It has random pits and bumps, but they appear to be minor imperfections on an approximately straight edge. You can easily spot the dominant trend. This is mild variation. By contrast, consider the rugged coastline of Brittany: Does it really have an "average" outline, like that of the razor blade? Only from the very great height of a satellite, where the familiar map shape can be imagined; but from closer up, in an airplane or from a tower, the tortuous, random details of promontories and bays, crags and hollows obscure the image. This coastline is wild. Yet a third example, this time in electronics. If you run a steady electrical current through a copper wire, you can "hear" it on a loudspeaker as a steady, white noise—the static of mild variation, due to the thermal excitation of the electrons. But if you try to run computer data down a very long wire, you will pick up irregular, intermittent "pops" and crackles on the line. Engineers call this 1/f noise, and it is the bane of computer communications, causing transmission errors. It cannot be predicted or prevented; it can only be accommodated, with error-correcting software. That is wild variation.

Wild randomness is uncomfortable. Its mathematics is unfamiliar and in many cases remains to be developed. It looks difficult, often requiring elaborate computer simulations rather than a quick punch on a calculator. Unfortunately, the world has not been designed for the convenience of mathematicians. There is much in economics that is best described by this wilder, unpleasant form of randomness—perhaps because economics is about not just the physics of wheat, weather, and crop yields, but also the mercurial moods and unmeasurable anticipations of wheat farmers, traders, bakers, and consumers.

This makes for strange conundra. Suppose you are asked to calculate the average size of companies in the software industry. So you go down a list, counting the firms, adding up their reported revenues, and dividing one number into the other to get a simple average. But how long should the list be? Just the top fifty publicly traded firms? Every company in an industry directory? Every firm that files a tax return and says it is in software? Impossible to say: Each time you lengthen the list and add more, smaller firms, your calculated average drops. And what about Microsoft? It is the colossus of the industry, dwarfing every other firm. Try to survey the industry: If you include Microsoft in the sample, it grotesquely inflates what the survey suggests the "typical" company value is. But if you exclude it, you ignore the most important company in the industry. In short, the distribution of company size is wild—Wild West, in the view of Microsoft's critics.

Back to Finance

Now, having wandered rather far, we come to financial markets. Suppose you can simulate on your computer an artificial stock market. Based on your ideas about how one piece of the economy connects to another, you build an elaborate econometric model. It inputs data you give it about the weather, population, inflation, economic growth, industry specialization, and the companies being traded; and it calculates what its algorithms tell it is the optimal, "fundamental" value of a company's stock. To that value, it adds millions of small random changes—perhaps reflecting fictitious news events, or fickle investor preferences, some taken separately, some added together. So, what kind of variability generator will you use? If mild, the resulting price charts will vary within a certain well-defined range; their trace will be the product of many small computer-generated events. Very different is wild variability, even though it can be "tuned down" to be less extreme than Cauchy's. Wild price charts will be a hair-raising record, mixing small movements with very, very large dislocations, many computer-generated news items with a few cataclysmic bulletins, many small transactions with large, institutional block trades—all, a mix of the small and routine with the large and rare. In such a wild world, an imaginary investor participating in this econometric simulation could be wiped out overnight.

Alas, this is not a computer fantasy. Hitherto, standard financial theory has followed the first, mild path. How it got on that mistaken path, and how it can get off it, will be seen in subsequent chapters.

Chapter III

Bachelier and His Legacy

In March 1900, the academic equivalent of a trial by fire was convened at the University of Paris.

The judges included Henri Poincaré, one of the most celebrated mathematicians of all time. He was a genius whose restless energy had led him across virtually every field of mathematical inquiry and beyond: probability, function theory, topology, geometry, optics, and, above all, celestial mechanics. He was a widely read popularizer of math and science, and his collected columns fill several books read to this day.

He was, however, a living paradox, both establishment figure and academic maverick. He was flippant, in the view of some colleagues, in his disregard for the theoretical niceties of mathematics; and he was relegated to various chairs of practical math where he could not corrupt the students' faith in perfect rigor. He played the absent-minded professor. Friends joked he was ambidextrous: equally awkward with either hand. His cousin, Raymond, became president of France during the First World War, then prime minister in the 1920s. And had Henri not died prematurely in his fifties, he would probably have received the Nobel Prize in physics (there is none in math). He had a keen sense of the beautiful in mathematics. He once said: "A scientist worthy of the name, above all a mathematician, experiences in his work the same impression as an artist; his pleasure is as great and of the same nature."

Before Poincaré on that day in 1900 was one of his doctoral students, Louis Bachelier.[1] Jobs for Ph.D.'s were scarce; and so the award of a doctorate in France was a formal, trying process. The young mathematician's schooling had been mediocre, at best. Now he had to pass two final tests before Poincaré and the doctoral "jury." The lesser one was an oral examination on a standard topic, chosen and approved beforehand. Bachelier's was on fluid mechanics; and it tested both his knowledge and oratory—an important consideration for a man who hoped to become a professor. The subject was a specialty of one of Poincaré's fellow judges, Joseph Boussinesq; so it cannot have been an easy test for Bachelier. But, according to the panel's final report, he demonstrated that it was a topic he "grasped deeply."

The main test was the defense of his original research; and the subject was not calculated to win easy approval. His thesis, "Théorie de la Spéculation," was not on complex numbers, function theory, differential equations, or other topics then in mathematical vogue. Nor was the *spéculation* to which it referred some form of speculative thought—no polite monograph on the philosophy of chance, this. It was about the money-grubbing form of speculation, the trading of government bonds on the Paris exchange, or Bourse, a thriving den of capitalism modeled after a Greek temple and located on the opposite river bank, geographically and intellectually, from the famed Sorbonne. Then as now in France, unbridled speculation had an unsavory reputation. While investment was socially desirable, pure gaming, or *agiotage*, was not. Futures trading on the exchange had been legalized only fifteen years earlier. And "shorting"—selling securities with borrowed money, to profit from a falling price—

was beyond the pale. While there had been some books on financial markets by 1900, its study was not yet an academic discipline, much less an appropriate topic for a provincial seeking approval and patronage from the great *Faculté des Sciences de l'Université de Paris*.

The professors were underwhelmed. Poincaré, reporting on the dissertation, observed that "the subject chosen by M. Bachelier is a bit distant from those usually treated by our candidates." He praised some of the "original" insights in the thesis, and suggested the most unusual one should have been more fully developed. But it was not the kind of paper that won the highest honors: The grade was a respectable *"mention honorable,"* not the *"très honorable"* that would have assured Bachelier a first-class ticket to an august mathematical career. And so he spent the next twenty-seven years battling for recognition and tenure from the French academic establishment. He shuttled across France as high school teacher and adjunct lecturer at Paris, at Besançon near the Swiss border, at Dijon and at Rennes. Fortunately for him the thesis appeared in a major journal and was not lost to history.

That history would judge him more kindly than did his contemporaries. His thesis laid the foundations of financial theory and, far more generally, of the theory of all forms of probabilistic change in continuous time. He formulated the basic questions of how prices move and proposed preliminary answers to some. He died unknown; more than a half century passed before his thesis was rediscovered and translated. On his ideas economists built an elaborate and comprehensive theory of markets, investing, and finance— how prices vary, how investors think, how to manage money, and how to define risk, the restless soul of the market. Their teachings found willing students on Wall Street and became the catechism for what is now called the "modern" theory of finance. Like any dogma, it is honored far more in the breach than in the observance; and most professional financiers and investment advisers, working from experience and intuition, have modified the specific formulae, as needs suit. But Bachelier's broad principles remain the framework on which much of the world's money is presented as moving.

Bachelier was "so outstanding in his work that we can say that the study of speculative prices has its moment of glory at its moment of conception," wrote a later economist, Paul H. Cootner of MIT. I, too, view Bachelier as a major figure in science. But at a very early stage, I proposed alternative theories of market dynamics, and as time passed devised better replacements.

An interest in the history of ideas is good for the scientist's soul. Hence, my books often devote a portion to the contemplation of individual scientists, physical or social, and their vicissitudes. To understand why the orthodox theory of financial markets and investment is so flawed, it first helps to review it—and there is no better way than by portraying a few men of the twentieth century who stand out as especially influential, regardless of whether one agrees with them or not. They are Louis Bachelier, Harry Markowitz, William Sharpe, and the duo of Fischer Black and Myron Scholes. The first, hero of this chapter, was a maverick, a lone visionary who overcame the general apathy and occasional opprobrium of his contemporaries and doggedly pursued his unique view of the financial world. The others, appearing in the next chapter, were secure in their professions and honored by their peers; their importance was to have made the boldest strokes that completed the canvas begun by Bachelier. There were many other hands, some of which historians might argue were equally significant. But every story must start somewhere, and this one must begin with Bachelier.

"Not an Eagle"

In 1900 as now, French academia was a cliquish, elitist institution, in which outsiders, dissenters, and mavericks were poorly tolerated. And Louis Jean-Baptiste Alphonse Bachelier was an outsider from the start.

He was born in the bustling port city of Le Havre on March 11, 1870, to a prosperous business family. His father, Alphonse Bachelier, was a wine merchant of high repute; the Venezuelan government named him its vice consul, or representative at the port. His grandfather was a local banker and minor poet. Louis grew into a comely young man, five foot nine, with blond hair, blue eyes and an aquiline nose, according to his military records. But his comfortable start in life was disrupted in 1889 by the death of both parents. Bachelier, then nineteen, stopped his schooling to work in the family business. Shortly after, he was drafted. That meant he missed the conventional path to academic security in France: a successful passage through one of the Republic's *grandes écoles*, where the nation's elite is formed and confirmed—the French counterpart to Britain's "Oxbridge" and the American Ivy League. In fact, he did not get back to school until he was twenty-two. He enrolled as a mathematics student in the University of Paris, which was open to all high school graduates. Lacking the preparation of his peers, he achieved only mediocre grades: One examination required more than one attempt to pass, and even then, he barely made it.

Bachelier did not always help his own career. He appears, in his writings and in the recollections of contemporaries, to have been a difficult man. Certainly, modesty was not a conspicuous virtue. In a 1921 curriculum vitae for a job application, he described his by-then voluminous writings, including two books and journal articles, as no mere academic scribbling; they were nothing less than "the renewal of a science that, born in France, had become the exclusive property of the Germans and English." His 526-page book on probability "surpassed the great treatise of Laplace." And he described another work as "absolutely unique to the author; he got the original idea from no one; no other work of the same type has ever been done. Conception, method, results—all are new."

His work, indeed, was original. But his contemporaries were not impressed. At lectures, students clucked that he did not complete equations on the blackboard without reference to his notes, hallmark of a great pedagogue. On one occasion, he admonished a class that it must know the Greek alphabet perfectly—and then proceeded to forget it himself. A 1921 letter by an official in the education ministry called him "a malcontent." It said he got one of his first teaching jobs at the intercession of the ministry, in recognition of his service as a lieutenant in World War I, but over the objections of other mathematicians. He clearly lacked the political finesse required to get ahead. "He is not an eagle," the functionary wryly observed.

The shortage of university chairs in mathematics at the time meant that, on the rare occasions a vacancy occurred, a scramble ensued. In 1926 such an opening arose in Dijon, where Bachelier had worked before. His rival for the chair, Georges Cerf, was a brilliant young mathematician with all the right connections in Paris and an ally in Dijon, Maurice Gevrey, a sitting math professor. Gevrey appears to have taken a passionate dislike to Bachelier. Scouring the latter's work, Gevrey soon spotted a glaring mathematical error. When the academic committee met to decide the professorship, Gevrey brandished a letter from the eminent French probabilist, Paul Lévy in Paris, confirming the fault. Result: "Bachelier was blackballed," as Lévy ruefully recalled years later, in correspondence with me. By then, Lévy regretted the incident. He had read only the passage highlighted by Gevrey rather than the entire treatise; and in the full context of Bachelier's work the error appears benign. Lévy later apologized to Bachelier that "an impression, produced by a single initial error, should have kept me from going on with my reading of a work in which there were so many interesting ideas."

Apologies were too late, however. Bachelier's response had been immediate and intemperate. He circulated a letter (of which several copies survive) exposing the "deplorable and iniquitous" manner in which his career had been sabotaged at Dijon. His rival, he whined, was eighteen years his junior "and did not serve in the war, when at 48 I was on the front as an officer." Gevrey's partisan conduct, he wrote, "would not astonish anyone who knew the weakness of his character." Another of the committee members, he sneered, "is well known for his ingenuity: He was able to make a vacuum of his physics course" by boring the students away. And he let fly a fusillade of more than four hundred vitriolic words at Lévy. He called Lévy's critique "violent and unjustified" and based on total ignorance of his work. The Parisian, who had just finished a book on probability, had not even bothered "opening my book" on the subject before writing his own, Bachelier complained. He concluded with an insinuation typical of the time: "Without doubt, it is inconceivable that M. Paul Lévy had wanted, by a sort of last-minute trick, to favor *un coreligionnaire*." Lévy was a Jew.

Given Bachelier's temper, it is remarkable that he ever won the security of a professorial chair—which he ultimately did, at Besançon. But that was twenty-seven years after his doctoral thesis, the work for which he is so well remembered today.

The Coin-Tossing View of Finance

The Bourse, the bustling Paris exchange, was at that time a world capital of bond trading. After the French Revolution, the government made restitution to some of the returning nobility by issuing a billion francs in perpetual bonds—a certificate that pays fixed interest indefinitely but never repays the principal. These *rentes*, as they were called, became a financial hit, widely held and actively traded: By 1900, 70 billion francs of domestic and international bonds were outstanding, compared to a government budget of four billion francs. As with U.S. Treasury bonds and U.K. Gilts today, such was the depth of the French bond market that parallel trading developed in related futures, options, and other derivatives with such exotic jargon as "call o' more's," "put o' more's," "spreads," and "contangoes." Bachelier was intimately familiar with the arcana of these markets, and he devoted part of his sixty-eight-page thesis to a detailed description of them. Indeed, some historians suggest he may have worked at the exchange for a while. His goal was to develop formulae to price these complicated derivatives. For that, he needed first to work out how the underlying bond prices themselves moved. Such "a formula which expresses the likelihood of a market fluctuation does not appear to have been published to date," he wrote. And for good reason:

> The factors that determine activity on the Exchange are innumerable, with events, current or expected, often bearing no apparent relation to price variation. Beside the somewhat natural causes for variation come artificial causes: The Exchange reacts to itself, and the current trading is a function, not only of prior trading, but also of its relationship to the rest of the market. The determination of this activity depends on an infinite number of factors: It is thus impossible to hope for mathematical forecasting. Contradictory opinions about these variations are so evenly divided that at the same instant buyers expect a rise and sellers a fall.
>
> The calculus of probability can doubtless never be applied to market activity, and the dynamics of the Exchange will never be an exact science. But it is possible to study mathematically the state of the market at a given instant—that is to say, to establish the laws of probability for price variation that the market at that instant dictates. If the market, in effect, does not predict its fluctuations, it does assess them as being more or less likely, and this likelihood can be evaluated mathematically
> *Opening lines of "Théorie de la Spéculation"*

Now, there had been a few earlier mathematical sorties at the market. A French stockbroker, Jules Regnault, had observed in 1863 that the longer you hold a security, the more you can win or lose on its price variations. He even worked out a formula for it. But most market analysis looked at stock and bond prices in the conventional way: Something happens and prices react—a story of cause and effect, easy to work out afterwards, difficult to forecast beforehand. But this approach was futile; one can never know everything. Instead, Bachelier tried to estimate the odds that prices will move—then, a novel approach. And he did so brilliantly by observing "a strange and unexpected" analogy between the "diffusion" of heat through a substance and how a bond price wanders up and down. Both, he saw, are processes that you cannot precisely forecast. At the level of particles in matter or of individuals in markets, the details are just too complicated; you can never discriminate and describe every relevant factor or analyze exactly how they all interrelate to spread energy or energize spreads. But in both fields, you can back away from the messy details of how or who and see the broad pattern of probability that describes the whole system. So, in the most specific of his models, Bachelier adapted the equations of one field to the problems of another.

In this model, he started by looking at the bond market as what he called "a fair game." Recall the old pastime of tossing a coin, discussed earlier. If the coin is fair, or unweighted, it is as likely to come up heads as tails on each toss. If you win a dollar for every head and lose a dollar for every tail, at the end of a string of tosses the language of probability says you should "expect" a profit of zero. Moreover, each time you toss the coin the odds of heads or tails remain 50-50, regardless of what happened on the prior toss. Put another way, a key idea behind a fair coin is that it has no memory: While you can get long runs of heads or tails, at each toss the run is as likely to end as to continue.

Now, God does not play games of chance with French bond prices; but it can certainly look that way to anybody in the thick of the trading, unable to see exactly what is driving the market up or down. And it can be described that way mathematically, with the resultant formulae used to make probabilistic statements about what could happen next. That was another of Bachelier's key insights. He assumed the split-brain thinking so common among economists today, two different ways of looking at the same event. One was after the fact or *ex post facto*, and the other before the fact or *ex ante*. After a price move, you can examine it and deduce a cause-and-effect "story" of why it happened; for instance, bond prices fell because of a new, gloomy inflation report, or because of new rumors that a big bond dealer was insolvent. But before the price move, it would have been difficult to predict those news events and even harder to forecast how the market would react. So, in your ignorance, you would have simply looked at the then-current bond prices and assumed they were fair, that the market had already taken account of all relevant information, and that prices were in equilibrium with supply matched to demand, and seller paired with buyer. Unless some new information came along to change that fine balance, you would have no reason to expect any change in price. The next move would as likely be up as down, left as right, north as south.

In effect, prices follow a random walk, the metaphor adopted by Bachelier's successors. The term comes from a quaint puzzler in probability theory. Suppose you see a blind drunk staggering across an open field. If you pass by again later on, how far will he have gotten? Well, he could go two steps left, three right, four backwards, and so on in an aimless, jagged path. On average—just as in the coin-toss game—he gets nowhere. So if you consider only that average, his random walk across the field will be forever stuck at his starting point. And that would be the best possible forecast of his future position at any time, if you had to make such a guess. The same reasoning applies to a bond price: In the absence of new information that might change the balance of supply and demand, what is the best possible forecast of the price tomorrow? Again, the price can go up or down, by big increments or small. But, with no new information to push the price decisively in one direction or another, the price on average will fluctuate around its starting point. So again, the best forecast is the price today. Moreover, each variation in price is unrelated to the last, and is generated by the same unchanging but mysterious process that drives the markets. The price-changes, in the language of statistics, form a sequence of independent and identically distributed random variables.

In fact, it is even simpler than that, Bachelier reasoned. If you plot all of a bond's price-changes over a month or a year onto a graph, they would spread out across the paper in the familiar bell-curve shape—the many small changes clustered in the center of the bell, the few big changes at the edges. This opened the whole kit of common mathematical tools for the normal, or Gaussian, distributions mentioned earlier. And thus, through the agency of Bachelier, Gauss's theoretical curve came to be applied to the analysis of financial markets.

But Bachelier also ventured into new mathematical territory. Nearly a century before, the great French mathematician Jean Baptiste Joseph Fourier had devised equations to describe the way heat spreads. Bachelier knew the formulae well from his physics lectures. He adapted them to calculate the probability of bond prices moving up or down, and called the technique "radiation of probability." Strangely, it worked. Also, as fate would have it, very different motivations had sent other scientists on this trail. Long before, the invention of the microscope led to observations of the erratic way that tiny pollen grains jiggled about in a sample of water. A Scottish botanist, Robert Brown, studied this motion, observed that it is not a manifestation of life but a physical phenomenon, and received (possibly inflated) credit for the discovery through the term "Brownian motion." In 1905, Albert Einstein developed for it equations very similar to Bachelier's own equations of bond-price probability—though Einstein never knew that. Regardless, one cannot help but marvel that the movement of security prices, the motion of molecules, and the diffusion of heat could all be of the same mathematical species. As will be seen, it is one of many such strange liaisons in nature.

Bachelier did not stop at theory: He also tested his equations against real prices for options and futures contracts. The theories worked. For instance, he calculated that the buyer of a forty-five-day option at half a franc has 40 percent odds of earning a profit. He was uncannily close: Looking back at real trading data, he found 39 percent of such options had in fact yielded a profit to their buyers. "The market, unwittingly, obeys a law which governs it, the law of probability," he concluded.

The Efficient Market

Alas, Bachelier's economic insights went largely unnoticed for many years. In those days, finance theory was an oxymoron; finance was a distasteful trade, not a subject fit for academic inquiry. That attitude did not start changing until the Crash of 1929. Then, more economists began trying to understand financial markets. Independently of Bachelier, some started to think about a random walk. Alfred Cowles III, a wealthy investor frustrated by the imprecision of what passed for financial advice, established a foundation to gather and analyze market data. In one 1933 paper, he found what Bachelier would have predicted: Among twenty-four stock-market forecasters whom Cowles systematically studied, he found "no evidence of skill." They might as well have been shooting craps. Twenty years later, a British statistician, Maurice G. Kendall, took a long look at London shares, New York cotton, and Chicago wheat—more than a century of data—in search of conventional patterns upon which an investor could turn an easy buck or quid. "On the whole," he laconically concluded after pages of fruitless regression analysis, "I regard this experiment as a failure. . . . There is no hope of being able to predict movements on the exchange."

But it was not until 1956 that Bachelier's name reappeared in economics, this time, as an acknowledged forerunner, in a thesis on options-pricing by a student of MIT economist Paul A. Samuelson. Bachelier's idea of a "fair game" caught on; and economists recognized the practical virtues of describing markets by the laws of chance and Brownian motion. They were, in the 1960s and 1970s, put into a broader theoretical framework by Eugene F. Fama. As a student at the University of Chicago, Fama contacted me at IBM and Harvard; I became his thesis adviser, by telephone, mail, and repeated visits. His dissertation was on my views of market dynamics (of which more, later). But we often discussed Bachelier's ideas beyond the model of independent increments, and in subsequent years Fama elaborated them into what is now called the Efficient Markets Hypothesis. It is the intellectual bedrock on which orthodox financial theory today sits.

At its heart: In an ideal market, security prices fully reflect all relevant information. A financial market is a fair game in which buyer balances seller. Given that, the price at any particular moment must be the "right" one. Buyer and seller may differ in opinion; one may be a bear, and another a bull. But they both agree on the price, or there would be no deal. Multiply this thinking by the millions of daily deals of a bustling market, and you conclude that the general market price must be "right," as well—that is, that the published price reflects the market's overall best guess, given the information to hand, of what a stock is likely to profit its owner. And if that is true—and here is the bitterest pill for an investor to swallow—then you cannot beat the market.

Consider three cases. First, suppose a clever chart-reader thinks he has spotted a pattern in the old price records—say, every January, stock prices tend to rise. Can he get rich on that information, by buying in December and selling in January? Answer: No. If the market is big and efficient then others will spot the trend, too, or at least spot his trading on it. Soon, as more traders anticipate the January rally, more people are buying in December—and then, to beat the trend for a December rally, in November. Eventually, the whole phenomenon is spread out over so many months that it ceases to be noticeable. The trend has vanished, killed by its very discovery. In fact, in 1976 some economists spotted just such a pattern of regular January rallies in the stocks of small companies. Many investors close their losing positions towards the end of the year so they can book the loss as a tax deduction—and the market rebounds when they reinvest early in the new tax year. The effect is most pronounced on small stocks, which are more sensitive to small money movements. Alas, before you rush out to trade on this trend, you should know that its discovery seems to have killed it. After all the academic hoopla over it, it no longer shows up as clearly in price charts.

Second case: Suppose a financial analyst, poring over France Telecom's annual reports and chatting with its bankers and competitors, concludes that the company's debt is getting too large. To keep paying it, it will have to cut its dividend, borrow more, or sell an important asset. Can the analyst get rich on that insight? Not if the market is efficient. Other analysts will swiftly spot the problem, too, and advise their clients to sell France Telecom short. Or the bankers, who reached the conclusion first and now fear a loan default, will start charging France Telecom extra for its routine credit lines. The market will notice that, and the stock will fall. Again, the information is quickly priced into the stock.

Third and final case: Suppose the France Telecom CEO starts cashing in his stock options, because he knows the debt is a time bomb. How long can he profit on his inside information? In an efficient market, not very long. Traders will notice the captain is abandoning ship, and figure something bad is about to happen. So they sell, too, and the stock falls.

That is the theory, anyway. In all three cases—reading price charts, analyzing public information, and acting on inside information—the market quickly discounts the new information that results. Prices rise or fall to reach a new equilibrium of buyer and seller; and the next price change is, once again, as likely to be up as down. That does not mean you cannot win, ever. In fact, by the simple odds of a fair game, you can expect to win half the time and lose half the time. And if you have special insights into a stock, you could profit from being the first in the market to act on it. But you cannot be sure you are right or first; after all, the market is full of people at least as smart as you. So, in sum, it may not be worth your while to spend all that time and money getting the information in the first place. Cheaper and safer to ride with the market. Buy a stock index fund. Relax. Be passive. Or as Samuelson at MIT put it: "They also serve who only sit and hold." His advice, then:

> A respect for evidence compels me to incline toward the hypothesis that most portfolio decision makers should go out of business—take up plumbing, teach Greek, or help produce the annual GNP by serving as corporate executives. Even if this advice to drop dead is good advice, it obviously is not counsel that will be eagerly followed. Few people will commit suicide without a push.
>
> *From* The Journal of Portfolio Management, *1974.*

A dark, nihilistic message. But then, Wall Street is nothing if not flexible—and so what could have been its epitaph was recast as a rallying cry. Bachelier's thesis was elaborated into a mature theory of how prices vary and how markets work. It came in the 1970s and 1980s to be the guiding principle for many of the standard tools of modern finance, the orthodox line, taught in business schools and shrink-wrapped into financial software packages, for how to value securities, how to build portfolios, how to measure financial performance, and how to judge the merits of a financial project. As will be seen, it is a house built on sand.

CHAPTER IV

The House of Modern Finance

IN 1999, two economists from Duke University engaged in what you might, at first glance, think a tedious bit of research. They surveyed the chief financial officers of the largest U.S. corporations to ask them how, exactly, they do their jobs. How do they decide which factories, acquisitions, or new ventures to fund, and which to kill? How do they determine whether it would be cheaper to issue stock, sell bonds, or just borrow from the bank? The questionnaire was three pages long and took about seventeen minutes to fill out; and seventeen minutes is rather a lot of time to ask of the paymasters of the universe that most big-company CFOs imagine themselves to be. Still, 392 responded.

The answer that came back: When it comes to estimating their cost of capital—an essential ingredient in any financial decision—the method used most widely was the Capital Asset Pricing Model, or CAPM. In all, 73.5 percent said they use it. Nor is this unique to U.S. Fortune 500 companies. A similar survey of CFOs in sixteen European countries in 2001 found the same acronym, CAPM, on the lips of 77 percent. It is also in the political phrase-books. When Central Hudson Gas & Electric Corp. wanted to raise its electricity prices in New York in 2001, CAPM was part of the rationale argued to the regulator. By contrast, when the utility regulator in Northern Ireland wanted instead to cut electricity prices in his jurisdiction, part of his rationale: CAPM. Clearly, a double-edged sword.

So what is this strange acronym? Why does it feed into so many financial decisions upon which prices, jobs, and mergers depend? More important: Are these decisions right? The professors, grading the questionnaires, tartly noted that the CFOs did not seem to be using the model correctly—or, at least, not the way it is taught in business schools. Most of them seemed to be using some other techniques, as well. And, the professors added: "Even if it is applied properly, it is not clear that the CAPM is a very good model."

But the model remains, they concluded, "widely used." Its origin goes back to Bachelier. CAPM is one way in which his theories have been translated into practical tools of finance. This particular one is a simple method for valuing an asset, whether a stock you may buy or a factory your company may build. It was devised by William F. Sharpe, an American economist, in the early 1960s. Another tool inspired by Bachelier is Modern Portfolio Theory, a method for selecting investments devised in the 1950s by Harry M. Markowitz, a University of Chicago Ph.D. A third: the Black-Scholes formula for valuing options contracts and assessing risk; its inventors were two East Coast researchers, Fischer Black and Myron S. Scholes, in the early 1970s. There are many others, some more recent; but these three innovations—CAPM, MPT, and Black-Scholes—are the most important elements of orthodox financial theory. They are part of the basic curriculum for nearly every MBA student in the world. They are on the American financial industry's exams to become a certified financial adviser. The precise extent to which they are actually used in real-world finance is unknowable and surely differs from firm to firm, and from task to task. Many have tried fixing flaws in the formulae, and have added new ideas. But these remain the principal building blocks with which the modern house of finance is constructed. And they all sit on the theoretical foundation laid by Bachelier a century ago.

This book argues that the foundation needs re-pouring, before any more repairs are done to the building. To understand why this matters, let us first look more closely at the structure as it exists today.

Markowitz: What Is Risk?

The first big step in applying Bachelier's ideas was taken by a Chicago grocer's son, Harry Markowitz. His was a comfortable background—perhaps not as wealthy as Bachelier's but, as he later recalled, "we lived in a nice apartment, always had enough to eat, and I had my own room. I was never aware of the Great Depression."

Still, it was the economics of uncertainty that most interested Markowitz when he got to the economics department at the University of Chicago. How do you decide to invest in a new factory or stock when you cannot know beforehand exactly how the investment will turn out? In the stock market, the conventional wisdom in Markowitz's day was simple: Become a good stock-picker, or hire one. Whether through experience, inside information, long research, or hard number-crunching, some people are just good at it. Of course, there were plenty of books on how to do it. You could analyze how much cash a company might generate, how much profit it would report, how much it had borrowed—and, massaging the numbers, come up with a guess for what each share, in an ideal world, "should" be worth. If the market price is lower, buy the stock. Eventually, the rest of the market will come to agree with you; the price will rise; and you will make a nice profit. If it sounds risky, no problem: Pick several stocks to spread your bets. If you are good at it, you will end up with enough winners to offset your losers. So went the theory.

Markowitz was studying in the university library in 1950 when his "eureka" moment came. The young economist was hunting a good topic for a doctoral thesis; and a chance meeting with a broker had led him to look at the stock market. "I had not taken any finance courses, nor did I own any securities," he recalled later. He had read some books, such as a 1934 classic on stock-picking, *Security Analysis* by Benjamin Graham and David L. Dodd. And he was in the library reading another, *Theory of Investment Value* by John Burr Williams. It argued that, to estimate a stock's value, you start by forecasting how much in dividends it will pay; then adjust the prediction for inflation, foregone interest, and other factors that make the forecast uncertain. A straightforward rule. But surely, Markowitz thought to himself, real investors do not think that way. They do not look only at their potential profit; if they did, most people would buy just one stock, their best pick, and wait for the winnings to roll in. Instead, people also think about diversification. They judge how risky a stock is, how much its price bounces around compared to other stocks. They think about risk as well as reward, fear as well as greed. They buy many stocks, not one. They build portfolios. "Don't put all your eggs in one basket": It was an idea as old as investing itself. Even Shakespeare knew it, as Markowitz later recalled:

> ...I thank my fortune for it,
> My ventures are not in one bottom trusted,
> Nor to one place; nor is my whole estate
> Upon the fortune of this present year;
> Therefore my merchandise makes me not sad.
>   Merchant of Venice, *Act I, Scene I*

So Markowitz pondered: How to translate those two concepts, risk and reward, into equations you can work with? Well, your hoped-for profit—the expected return—depends on what you think the stock price most likely will be when it comes time to sell.

"Most likely," if you go back to the familiar bell curve, would mean the average, or mean, of all the prices you expect it might hit before you sell. Risk is more difficult to define. Perhaps, he thought, risk depends on how much the stock price swings up or down around the mean—or, to put it another way, the odds that you guessed wrong about the final price. Again, back to the bell curve: The most common measures of volatility are called variance and standard deviation; the latter is just the square root of the former. So Markowitz went back to the bookshelves, and pulled down a 1937 textbook, *Introduction to Mathematical Probability* by J.V. Upensky. The math for it was already there, in black and white: "I saw these correlations pop out of the page. . . . I was elated." He worked up his ideas into a thesis. Of course, like Bachelier before him, Markowitz had some critics. At the defense of his doctorate, he later recalled, one of Chicago's best-known economics professors, Milton Friedman, argued that "they could not award me a Ph.D. degree in Economics for a dissertation which was not in Economics." Markowitz added that he assumed Friedman was "only half serious," as the degree was granted.

And his ideas spread. They had practical appeal. Markowitz was saying that the prospects for every stock can be described by just two numbers, the reward and the risk—or, mathematically speaking, the mean and the variance of what you expect the stock will pay back by the time you sell. You predict the first number, the average expected selling price, with standard stock-analyst tools: Make earnings forecasts, estimate dividend growth, or ask the chairman's bartender. You conclude, say, that in one year General Motors stock will be about 10 percent higher because you think that its earnings will grow nearly that much. You predict the second number, the variance, by using the bell curve as a yardstick when looking back at how the stock did in the past: On two-thirds of the trading days in the previous year, GM's stock moved by less than its 17 percent standard deviation, so chances are it will do the same in the next year. Conclusion: You expect GM stock will give you a 10 percent profit, with two-to-one odds that you will not be wrong by more than 17 percent; that is, you are unlikely to lose more than 7 percent, and you may make as much as 27 percent. That is precision—or the appearance of it, anyway. With it, you can systematically compare stock with stock, GM with Ford, IBM with GE. Plot them all on a graph. They will scatter across the page in a spectrum of mean and variance, profit and risk, risky losers in one corner and safe winners in another. Then the final step: Combine the stocks to build portfolios.

But how? Each possible combination of stocks will give a different overall return and a different overall risk—and estimating them is not a simple matter of adding all the numbers for all the stocks together. Portfolio risk is more complicated than that: The whole can be greater, or less, than the sum of the parts. Stocks have a tendency to move up and down together. If a recession is looming, many stocks across the whole market will start to fall; their movements are, to a greater or lesser extent, correlated. Markowitz likened it to the coin-tossing game. If you bet on one hundred fair coins and their tosses are all uncorrelated, you will probably come out even. The heads on some will counterbalance the tails on the others. Your bets are diversified. But it is bad news if the coins are correlated. That would be, he wrote, as if the coins "agreed among themselves to fall, heads or tails, exactly as the first coin falls." In an instant, depending on how the coins decide the toss, you are rich or broke. Stocks are a bit like that. Each stock, depending on the company sector or strategy, is correlated more or less with other stocks. So—and here comes the trick of Markowitz's portfolio theory—if you mix some stocks that flip tails with others that flip heads, you can lower the risk of your overall portfolio. If you do it right, you need not sacrifice too much profit. GM stock, with its heavy dependence on consumer spending, tends to rise when the economy booms and fall when it sours. Lilly stock is more stable—even, perhaps, a bit contrarian, because more people may get sick and buy medicine when the economy suffers. The two stocks are uncorrelated. So if you buy a bit of each, you will profit both when the economy grows and when it contracts.

Next step: Add more and more stocks in different combinations to build what Markowitz called an "efficient" portfolio. Efficient is a cheerful word put to many uses. A good pump is efficient; it moves the most water for the least energy. A portfolio is efficient if it produces the most profit with the least risk. Thus, with Markowitz's math, for each level of risk you contemplate, you can devise an efficient portfolio that will yield the highest possible profit. And for each level of profit you target, there is an efficient portfolio with the lowest possible risk. If you plot all these portfolios on a graph, they form a smooth, rising curve: go-go and risky portfolios towards the top, boring and safe ones down below.

So which do you buy? That depends on your appetite for risk. If you are greedy and fearless, go for a racy portfolio near the top of the graph. If you are timid, take a sleepy one near the bottom. The choice also depends on the rest of the economy—specifically, how attractive the stock market's main competitor, safe Treasury bills, appear to be. If interest rates are high and T-bills pay well, then you will not touch stocks unless you think they will pay more; but to get more, you may have to accept more risk. By contrast, if rates are low, a duller, safer stock portfolio may suffice. Another economist, James Tobin, saw portfolio-building as a two-step dance. First devise the optimum, efficient stock portfolio for the economic and market climate. Then decide how much of a gambler you are. If you are a risk-averse widow, leave most of your money in the bank and invest only a little in the efficient stock-market portfolio. If you are a typical investor, put at least half into the efficient portfolio and leave the rest in the bank. And if you are a high-roller with a death wish, put everything you have into the stocks—and then borrow to buy even more.

Thus, Markowitz and others transformed investing from a game of stock tips and hunches to an engineering of means, variances, and "risk aversion" indices. In fact, the term "financial engineering" has been popular on Wall Street ever since. There were problems, of course. First, as Markowitz himself pointed out, it is not certain that using the bell curve is the best way to measure stock-market risk; it is easy, but not necessarily right. Second, to build efficient portfolios you need good forecasts of earnings, share prices, and volatility for thousands of stocks. Otherwise, garbage in, garbage out. Finally, for each stock, you must laboriously calculate its "covariance" with, or how it fluctuates against, every other stock. For a thirty-stock portfolio, about the minimum needed to make the numbers work well, that means 495 different calculations of mean, variance, and covariance. For the entire New York Stock Exchange: 3.9 million calculations. And, because prices change, the exercise needs constant repetition. This was a daunting prospect, even for the expensive new IBM mainframe computers that in the 1960s were starting to appear on Wall Street.

Sharpe: What Is an Asset Worth?

- The answer to the number-crunching problem came first from a young economist who knocked on Markowitz's door one day in 1960.

William F. Sharpe knew a fair amount about the economics of uncertainty himself. He was born in Boston, the son of a Harvard University placement officer; but when World War II began he and his family migrated from city to city, following his father's military assignments. He also changed universities and concentrations— from the University of California in Berkeley to the campus in Los Angeles, and from medicine to business to economics. His thesis, on a classic economics topic called transfer pricing, was not going well; in fact, one professor advised he drop it. Another suggested he go visit Markowitz, who had left Chicago and was working near UCLA at a well-known think-tank, the RAND Corp. "I introduced myself to him and said I was a great fan of his work," Sharpe recalled later. And, of course, Markowitz had a good thesis idea, for which he became Sharpe's unofficial adviser: Simplify the portfolio model.

For that, Sharpe asked a question: What happens if everybody in the market plays by Markowitz's rules? The answer was surprising. There would be not as many efficient portfolios as people in the market, but just one for all. If fluctuations in stock prices suggested a second, better investment palette, then everybody would start moving their money into that new portfolio and abandoning the first. Soon, there would again be just one portfolio, the "market portfolio." So the market, itself, was doing the Markowitz calculations. It was the most powerful computer of all, producing tick-by-tick the optimum investment fund. Thus was born the notion of a stock-index fund: a big pool of money, from thousands of investors, holding shares in exactly the same proportion as the real market overall. Of course, the details are not so simple. First decide what you mean by "the market": just the thirty industrial stocks in the Dow, or the hundred shares in the British FTSE index? Should you include bonds? What about other risky assets, like home equity? And, whatever the market, you will still need to keep re-tuning the fund to track it. Buy or sell too much or at the wrong time, and you lose money.

But there is more. If all that matters is the market portfolio, then the value of an individual stock depends only on how it compares to the rest of the market. Of course, the performance of the market overall flows with the economic tides. Treasury bills are safe and dull; since 1926, they have paid an average 3.8 percent with very little risk in good times and bad. The stock market, by contrast, is risky and exciting; in the same period, the Standard & Poor's 500 stocks have paid an average 13 percent, but with huge swings around that average, through bear and bull, crash and boom. The gap between the average stock-market and T-bill profits is what economists call the equity risk premium. Think of it is as the price the stock market must pay to lure people's money away from safe banks and government bonds.

Now look at an individual stock. One that tracks the market—is exactly correlated with it—will pay no more nor less than the market overall. But a stock that plummets 4 percent when the market falls 2 percent is unattractive. It is twice as volatile as the market; you will not touch it unless you think that, despite the risk in bad times, it could still double your money in good times. Similarly, a stock that slips only 1 percent when the market falls 2 percent—or better still, rises when the market falls—may be very attractive. It is half as volatile as the market; you will pay more for it, and be content to make a smaller but safer profit. The amount by which the stock reacts to the market is the stock's "beta" or β, a Greek letter commonly used in mathematical equations. In plain English: To buy a stock, you have to expect it will pay you more than safe and sound T-bills. That "more" is proportional to how closely the stock mirrors the overall market's performance. Take an example. The stock of Hot TechCo may have a β of 1.5, which means it is very sensitive to the state of the market and economy. Plug the numbers into Sharpe's formula. Start with what T-bills are paying: 2 percent. Add to that another number, the stock's β (1.5) times the market's risk premium over T-bills (9 percent). What return can you expect it to pay? Answer: 2 percent plus (1.5 times 9 percent) equals 15.5 percent. That is a lot to expect a stock to pay you back in one year, but not impossible if you think the market is too gloomy about software stocks, and will eventually correct itself.

The concept is straightforward. It says the more you risk, the more you expect to get paid. It says the most important risk you face as a stock-market investor is the general state of the economy, reflected in how the market is doing. It says that if you are rational you would not normally want a stock that is going to die just as a recession arrives and you are about to get laid off; so to buy that stock, you have to be thinking it will rise so far in the good times that it will more than pay its losses in the bad times. And now a practical point, which helps explain why this formula became so popular in the world of finance. It takes all of Markowitz's tedious portfolio calculations and reduces them to just a few. Work up a forecast for the market overall, and then estimate the β for each stock you want to consider. From 495 calculations for a thirty-stock portfolio with Markowitz and portfolio theory, you simplify to thirty-one with Sharpe and the Capital Asset Pricing Model, as it came to be called. Looking at the entire New York Stock Exchange: From 3.9 million with Markowitz, you prune to 2,801 with Sharpe. This is no longer a job for a mainframe and a statistician; it is for a personal computer and a broker, or even an individual investor.

The impact of Sharpe's formula was not immediately apparent, even to him. After finishing his thesis, he wrote his ideas up in an article for publication in one of the leading academic journals—a process, as every researcher knows, fraught with uncertainty, politics, and, often, disappointment. Sharpe later recalled:

> I didn't know how important it (the CAPM paper) would be, but I figured it was probably more important than anything else I was likely to do. I had presented it at the University of Chicago in January 1962, and it had a good reaction there. They offered me a job. That was a good sign. I submitted the article to the *Journal of Finance* in 1962. It was rejected. Then I asked for another referee, and the journal changed editors. It was published in 1964. It came out and I figured OK, this is it. I'm waiting. I sat by the phone. The phone didn't ring. Weeks passed and months passed, and I thought, rats, this is almost certainly the best paper I'm ever going to write, and nobody cares. It was kind of disappointing. I just didn't realize how long it took people to read journals, so it was a while before reaction started coming in.
>
> *From an interview with* Dow Jones Asset Manager, *1998*

As luck would have it, others, including John Lintner, a Harvard professor, Jan Mossin, a Norwegian economist, and Jack Treynor, a consultant at Arthur D. Little Inc., were independently pursuing similar ideas. A very academic contest—in slow motion—took place. Sharpe was first to publish. But today most economists credit Sharpe, Lintner, and Mossin jointly for the development of the CAPM model.

Today, nearly every business school in the world teaches the model, and not just for valuing stocks. As mentioned earlier, the model turns out to be a handy tool for valuing a company's projects, too. Suppose you are an electronics CFO needing to decide whether to build a new semiconductor factory. You use CAPM to look at it from the point of view of an investor considering whether to include your shares in his portfolio—that is, to give you the cash to build the plant. Your company overall may have a moderately high β of 1.1; but the semiconductor industry may have an even higher β of 1.7. Thus, building the plant will expose your company to some extra risk. To justify that, the project's planners must show you that it can also produce extra profit. In short, CAPM helps set a "hurdle rate" for the plant's expected return. If the forecasts suggest the return will clear the hurdle, fine; build it. If not, scrap it. A similar rationale runs through the rate-making decisions of many utility regulators. To get money from the stock market, ConEd needs to offer investors a certain minimum return—and the precise amount depends partly on its β. If the company's projected profits are not high enough, the regulator may grant a rate increase to match the market's expectations. If too high, it will cut the rates. Of course, that is just the theory. In practice, a host of other assumptions feed into every CAPM calculation—and thus a seemingly objective CAPM answer can become quite as subjective as any other political process.

Black-Scholes: What Is Risk Worth?

The next big step in the development of modern financial theory began in a small, windowless smoking lounge in the Chicago Board of Trade. For more than a century, the exchange had been the center of U.S. trading in commodities—wheat, pork bellies, corn and soy, cattle and barley. But on April 26, 1973, some of its members inaugurated a new kind of market, in stock options.

Options of one kind or another had been around for generations; Bachelier's thesis had, after all, been on options. They gave the holder the right to buy or sell something at a fixed price. Stock options—contracts to buy or sell a company's stock—are a form of compensation for many executives, and for some speculators they are another way to bet on the stocks themselves. As an example, consider some of the options traded on the first day of trading in Chicago in 1973: contracts to buy 100 Xerox shares at $160 apiece over the following three months. Thirty-nine such contracts were traded. Each contract cost the buyer $5.50 a share. In New York that day, the actual Xerox stock price was only $149—meaning that the buyer of these $160 "call" options was betting Xerox would rise fairly quickly. If it hit, say, $170 within the next three months, he had the right to buy the stock cheaply at $160 and resell it at $170. Profit: $10 a share, less commission, taxes, and the original $5.50 option premium. On the other hand, if Xerox did not rise—or even fell—then the option would expire unused and the $5.50 premium would be lost. As fate would have it, that is exactly what happened with those first Xerox contracts in April 1973: They expired worthless because the stock did not rise high enough.

Before the new Chicago market, trading options was a small, expensive business conducted "over-the-counter," broker to broker, by phone or telex. The new market was an open bazaar, with published prices and low commissions. So for a speculator, those Xerox options were a new, cheaper way to bet on Xerox stock. The entrance fee was only $5.50 a share in Chicago, compared to $149 in New York. And it was safer: Even if the bottom fell out of Xerox, the option holder could not lose more than the initial $5.50 premium. But how was that premium set? As in any market, it was not dictated by anyone; it was just the value at which a buyer and a seller came together. But was there a way to estimate a "reasonable" price?

The answer came, not in the noisy Chicago trading pit, but from a mindset far away in Cambridge, Massachusetts. Fischer Black had started with a conventional academic career, graduating from Harvard in physics and then getting a doctorate in applied mathematics. He was a tall, thin man of few words. He complained of a poor memory, and so got into the habit of jotting down his ideas immediately, whenever or wherever they struck. As a lecturer, he was known for sometimes stopping in mid-sentence, falling silent, and taking notes. In 1965 he left Harvard and moved across town to a big Cambridge consulting firm, Arthur D. Little Inc.; he wanted, he later said, to work on practical problems with "more immediate payoff." There, he met another ADL man, Jack Treynor—the same who had devised, but not published, an asset-pricing model around the same time as Sharpe. Black began to study it, and became hooked. "The notion of equilibrium in the market for risky assets had great beauty for me," he recalled. He tried applying the model beyond stocks, to bonds, cash, and finally, to warrants, a close cousin of options.

Now, many smart people before Black had tried to find a formula for valuing warrants or options—including Bachelier and Paul Samuelson, the MIT economist. One common problem was that, to figure out what an option or warrant was worth today, they thought they had to know what the underlying stock would be worth at expiration—that is, how far "in the money" or "out of the money" the option would end up being. But that was a hopeless approach. If you could predict that, you would not be a struggling young economist much longer. As Black thought about it, he realized that maybe he could work around not knowing the stock's final value. He devised a complicated differential equation to describe his ideas—and then could not solve it; that type of math had not been one of his strong points. "So I put the problem aside and worked on other things," Black later recalled.

About this time, a young Canadian economist, Myron S. Scholes, arrived at MIT to start teaching finance at its Sloan School of

Management. Scholes had been born in a gold-mining region of the far north, Timmons, Ontario, where his father had moved in the Depression to set up a dental practice. His mother died of cancer when he was sixteen, and then he developed scar tissue on his corneas that, until corrected surgically ten years later, made it hard for him to read. "Out of necessity I became a good listener," Scholes later recalled. "I learned to think abstractly and to conceptualize the solution to problems." He went on to get a doctorate in economics from the University of Chicago, and was then offered a teaching job at MIT. There, several smart young economists had gathered around economists Paul Samuelson, Franco Modigliani, and Paul Cootner (the first two eventually won Nobels). And on Tuesday evenings, a workshop on finance met to discuss new issues. There, Scholes and Black got to know each other. Together, they took up Black's work again. They made an odd couple—the austere, reserved Harvard man and the temperamental, disputatious Canadian.

They focused on Black's earlier, counterintuitive insight: When valuing an option, you do not need to know how the game will end—that is, what the stock price will finally be when the option expires. Instead, all you need to know is what the traders themselves know, the terms of the option (the strike price and time to expiration) and how volatile the stock is. If a stock is very stable, its out-of-the-money options will not be worth much to anybody. The odds are very low that the stock price will rise far enough to make the options useful. By contrast, if a stock is risky, if its price swings widely up or down, then the options will be very valuable: Odds are high that, on one of those swings, the options will come into the money and pay off handsomely. Moreover, as the option matures and the stock price moves, the value of the option in the market-place will keep changing. The Black-Scholes formula permitted the same, frequent recalculations of value that the market itself did. It also, to be manageable, had "to assume away all kinds of complications," Black later recalled. For instance, they followed Markowitz,

Sharpe, and Bachelier in assuming that a stock's risk, or volatility, can be gauged by the bell-curve standard.

Black and Scholes started talking their ideas out with another MIT colleague, Robert C. Merton. He was a Columbia engineering undergraduate, a Caltech masters student in applied mathematics, and then an MIT economist—and he was bit badly by the markets bug. As a grad student, he would rise early to get down to a local brokerage house for the opening of trading; only after a few hours of watching the tape and placing his bets would he go to class. Now at MIT, as an assistant to Samuelson, Merton was also working on the options problem, and made some useful mathematical suggestions to Black and Scholes. But with Merton, the other two were both rivals and colleagues, so the collaboration was not complete. Merton missed his colleagues' first formal presentation of their equations, at a conference in Cambridge. He overslept.

Black and Scholes did not stop with the theory: They also tried it out, literally. They started with warrants and noticed several in the market that, according to their formula, looked cheap. The best were in a company called National General.

> Scholes, Merton and I (Black) and others jumped right in and bought a bunch of these warrants. For a time, it looked as if we had done just the right thing. Then a company called American Financial announced a tender offer for National General shares. . . . (That) had the effect of sharply reducing the value of the warrants.
>
> *"How we came up with the option formula," Black 1989*

In other words, they lost their shirts. But they did not care. The fact that their formula had correctly spotted the anomalous warrants suggested that their math was sound, even if their market intelligence was not. So in October 1970, Black and Scholes submitted a paper to the *Journal of Political Economy*. Rejection: Too specialized, the journal said. They tried another journal.

Rejection: Too many papers competing for too little space in its pages. Black, himself, suspected the ivory-tower class system at work. He later grumbled, "One reason these journals didn't take the paper seriously was my non-academic return address." In the end, the paper was rewritten and published in the *Journal of Political Economy*—but only after two friends from the University of Chicago, Fama and Merton Miller, lobbied the journal's editors to give it a second look.

Their article appeared in print just after the opening bell on the Chicago Board Options Exchange in 1973. It met an eager audience. Within a few years, options dealers had incorporated its esoteric terminology, of "deltas" and "implied volatilities," into their daily language. Texas Instruments began advertising its latest calculator as just the thing for a quick Black-Scholes calculation on the fly. An entire industry grew. With the help of the Black-Scholes formula and its many subsequent amendments, corporate financiers now routinely buy insurance, or hedge, against unwanted market problems, and not just in stocks. For instance, when General Electric signs a contract to deliver turbines to a British electricity company, it will buy pound "put" options whose value will rise if the pound falls. Similarly, fund managers can try to take out portfolio insurance—buying stock options that will zig when their portfolios zag. Certainly, these are costly; but they are cheaper than watching a portfolio shrivel when the market turns against it. And such hedging, or insurance, is the least of Black-Scholes's uses. Thousands of business executives find it in their pay: the formula is routinely used to calculate the value of the stock options a company grants its leaders. And it has permitted an entirely new type of trading, not in stocks or currencies themselves, but in their volatility. Traders can construct elaborate combinations of options so they cash in not at a specific price, but when prices swing more wildly, up or down, than normal. Or, they can do the opposite: Design an options package that pays off only if prices are steady. In that sense, the formula puts a price on risk.

Spreading the Word on Wall Street

Black returned to academia for a while, teaching finance at MIT's Sloan School. There he structured a popular course around a simple theme, fifty questions on finance. Alas, by then, many an experienced financier or economist would have skipped the class; the answers were starting to become clear, they thought.

Black's ideas and those of other theorists had already become dogma in the financial industry. They filled a need. The 1950s and 1960s had been a time of easy living on Wall Street. Most stocks rose with the postwar boom, and a broker's job was to deliver good "picks." Usually, that meant touting "growth" stocks like Xerox, IBM, or Avon, all members of a group of fifty well-reputed, fast-growing companies called the "Nifty Fifty." But in the 1970s, inflation and economic turmoil ended all that: The bear market of 1973–1974 wiped 43 percent off stock values, and the end of the gold standard for the dollar turned the sleepy currency market into the world's biggest casino. Then the options market, initially dismissed as an eccentricity of some hyperactive Chicago traders, turned out to be a major new branch of finance. The financial industry needed new tools, new answers. Sure, the academics and their theories were grumpy and difficult, and their message that you cannot beat the market was particularly galling. But Wall Street's customers were even more grumpy and difficult.

So the financial industry became a convert to the new, "modern" finance. Merrill Lynch turned CAPM into an industry in its own right, producing a periodical "Beta Book" for its brokers and customers eager to do the math themselves. Across the world, financial firms started constructing efficient portfolios for their clients. After a few false starts, the index fund, the ultimate in passive investing, was born. It now constitutes more than a fourth of U.S. fund investments. Options took off. The industry was transformed. It discovered economies of scale: If there is just one market portfolio and one

size fits all, then the same funds and same analysts can serve all customers. Merge and save. Bigger is better. And the academics themselves turned from disparaged outsiders to valued insiders. Many
joined or became consultants to big financial houses. Some—
Sharpe, Markowitz, Scholes, and Merton—received Nobels.

The whole edifice hung together—provided you assume
Bachelier and his latter-day disciples are correct. Variance and standard deviation are good proxies for risk, as Markowitz posited—
provided the bell curve correctly describes how prices move.
Sharpe's beta and cost-of-capital estimates make sense—provided
Markowitz is right and, in turn, Bachelier is right. And Black-
Scholes is right—again, provided you assume the bell curve is relevant and that prices move continuously. Taken together, this
intellectual edifice is an extraordinary testament to human ingenuity. But the whole is no stronger than its weakest member.

The crash of October 19, 1987, took many by surprise. On one
day, the Dow plunged 29.2 percent. Something was wrong: The
academics said that the fall should not have happened, that it was a
once-in-an-eon event. The carefully designed investment portfolios
blew up. The options-based portfolio insurance failed—indeed, it
made the market rout worse, as fund managers rushed to get more
insurance and thereby drove down prices even further. Later, the
financial turmoil of the 1990s reinforced the point: Something is not
quite right in the theory.

As the old saying goes, a fool and his money are soon parted. But
Wall Street is more accustomed to being parter than partee. And so
began a search for new ideas. It continues to this day. The old models are still taught, refined, retailed, and used, but they are no longer
viewed with quite the same degree of respect. As will be seen, that is
just as well.

CHAPTER V

The Case Against
the Modern Theory
of Finance

IF MONEY IS AN IDOL, then one of the largest temple compounds of this modern faith sits on a tight bend of the River
Thames, a few miles downstream from central London. There, in
the Canary Wharf business district, rise eighteen steel and glass
towers to which, each working day, 55,000 people commute to play
their part in the international money market. These are the inheri-
tors of Bachelier, Markowitz, Sharpe, Black, Scholes, and others:
fund managers who balance risk and reward, bankers who calculate
default risks, currency traders who place elaborate bets on options.
Their collective brainpower, both carbon- and silicon-based, is
astounding. As an industry, finance buys more computers than
almost any other. It hires a huge proportion of the world's newly
minted math and economics graduates. It is a vast calculating
machine, a robot to hang an electronic price tag on every product,
service, company, and country that deals in global commerce.

This is where financial theory, from Bachelier to Black, meets
financial reality. All the academic models are here, in the computers

and workbooks of the pros—but almost invariably updated, altered, or mixed with other models. Indeed, the result is something like a traditional medicine or over-the-counter nostrum: many different chemicals and no clear "active ingredient." But in the world of finance, the purity or elegance of the theory does not matter. Only one question counts, what makes money? And there are no easy answers.

Indeed, in the eyes of the academic purists, you would find lots of things that look plain wrong on a typical, real-world trading floor—so many that, when visiting one, you can play the old childhood game of "spot the mistakes" in an intricate picture.

Citigroup runs one of the biggest foreign-exchange operations at Canary Wharf. On a typical day in 2003, it is crowded, busy and self-absorbed. The Citigroup trading room is vast, with hundreds of computers, ceilings, track lighting, and 130 currency traders and salespeople arrayed along rows of desks, six to a side. Above the desks, small flags—the Union Jack, the Stars and Stripes, the Rising Sun—mark the currencies in which each cluster of traders specializes. Their language is colorful and arcane: "Nokie-Stokie" for trades between Norwegian and Swedish kronor (Nokie for the currency's computer code, NOK; Stokie for the Swedish capital, Stockholm); "cables" for the dollar-pound market whose rates were once cabled across the Atlantic; "plain vanilla" for the most common, standardized currency options. Each day, the multinational bank moves about one-ninth of all the world's internationally traded dollars, yen, euros, pounds, zlotys, and pesos; and about a third of its global "FX" business happens on the second floor of the London office.

But consider the "mistakes" on this floor. Seated at one row of desks, a pair of analysts spend their days studying the orders of the bank's own customers. They are looking for broad patterns they can report back to the clients in regular newsletters. Theirs is the sort of market-insider information that, one form of the Efficient Market Hypothesis says, should not be useful; any profitable insights into trading data should already be reflected in the prices. But they do not buy that notion: "The biggest edge you can have is the private information of who's buying what," says one of the analysts. "We do not believe the market is efficient."

Second mistake: A few desks down is a math Ph.D. from Cambridge. He spends much of each day studying the fast-changing "volatility surface" of the options market—an imaginary 3-D graph of how price fluctuations widen and narrow as the terms of each option contract vary. By the Black-Scholes formula, there should be nothing of interest in such a surface; it should be flat as a pancake. In fact it is a wild, complex shape. Tracking it and predicting its next changes are fundamental ways in which Citigroup's options traders make money. About 10 percent of the world FX options market is of a class called exotic. It has mind-numbing combinations of precise options terms tailor-made to pay off only under certain circumstances. These combinations are obscure to most people, but perhaps just what the CFO of GM needs to guard against one particular risk that worries him in his company's yen-based cash-flow. None of this would exist if the original Black-Scholes formula were accurate. Of course, the formula remains important; it is the benchmark to which everyone in the market refers, much the way, say, people talk about the temperature in winter even though whether they actually feel cold also depends on the wind, the snow, the clouds, their clothing, and their health. Citigroup's options analysts have the Black-Scholes formula in front of them all the time, in spreadsheets. But it is just a starting point.

Third mistake: the research department. Now, by orthodox theory, there should be no research department. You cannot beat the market, so all you need are a few traders and computers to stay even with it. But Jessica James, a Citigroup research vice president, punches up on her computer screen a simple chart, a graph of the dollar-yen exchange rate over the past decade. It wiggles across the screen, a seeming random walk reflecting the world's mercurial views on the relative merits of the American and Japanese economies: up, down, or sideways in what the eye sees as an irregular pattern, but which standard financial theory calls random fluctuation. Then she performs an elementary task, of the sort chartists have been doing for a century. She calculates a moving average—for each day, the average of the exchange rate over the prior sixty-nine days. This calculation traces a smoother, gentler line than the raw price data, averaging out all the peaks and troughs. Now, she suggests, here is a simple way to make some money in the currency market: Every time the actual exchange rate climbs above the average line, you buy. Every time it falls below the average line, you sell. Simple.

The result? If you had followed this strategy over the past decade, she calculates, you could have pocketed an average annual return of 7.97 percent. Heresy. Impossible. According to the Efficient Market Hypothesis, there should be no such predictable trends. Certainly, skepticism is warranted. As James notes, there is a big difference between spotting veins of gold in old price charts and minting real gold in live markets. Those 7.97 percent average returns included some periods of hair-raising loss, when sticking to the strategy would have required steel nerves and deep pockets. Still, a by-now substantial body of economics research suggests that there is, indeed, money to be made in such a "trend-following" strategy; how much, and whether it is worth the risk and expense, is a matter of debate. But clearly, the market pros have already voted: More than half of currency speculators play some form of trend-following game, market analysts estimate.

So how to explain so stark a discrepancy between theory and reality? Start by looking at the assumptions underpinning the theory.

Shaky Assumptions

All models by necessity distort reality in one way or another. A sculptor, when modeling in stone or clay, does not try to clone Nature; he highlights some things, ignores others, idealizes or abstracts some more, to achieve an effect. Different sculptors will seek different effects. Likewise, a scientist must necessarily pick and choose among various aspects of reality to incorporate into a model. An economist makes assumptions about how markets work, how businesses operate, how people make financial decisions. Any one of these assumptions, considered alone, is absurd. There is a rich vein of jokes about economists and their assumptions. Take the old one about the engineer, the physicist, and the economist. They find themselves shipwrecked on a desert island with nothing to eat but a sealed can of beans. How to get at them? The engineer proposes breaking the can open with a rock. The physicist suggests heating the can in the sun, until it bursts. The economist's approach: "First, assume we have a can opener. . . ."

The assumptions of orthodox financial theory are at least as absurd, if viewed in isolation. Consider a few:

1) Assumption: People are rational and aim only to get rich.

Theory:

When presented with all the relevant information about a stock or a bond, individual investors can and will make the obvious rational choice that leads to the greatest possible wealth and happiness. They will not ignore important information, or pay a lot for a stock they expect to fall. They will not become philanthropists. They will behave as rational, clear-thinking, self-interested individuals, each one a latter-day Adam Smith. They will make the market work efficiently, with their well-reasoned actions driving prices quickly to the "correct" level. And their preferences can be expressed in straightforward formulae, economic "utility functions" that, for a given input, always yield the same output. In the language of economics: The greatest wealth and happiness maximize utility. In short, rational investors make a rational model of the market.

Reality:

People simply do not think in terms of some theoretical utility measurable in dollars and cents, and are not always rational and self-interested. The refutation of this one assumption of modern financial theory has in the past twenty-five years created a fertile new field of inquiry, called behavioral economics. It studies how people misinterpret information, how their emotions distort their decisions, and how they miscalculate probabilities. For instance, suppose you offer somebody a choice: They can flip a coin to win $200 for heads and nothing for tails, or they can skip the toss and collect $100 immediately. Most people, researchers have found, will take the sure thing. Now alter the game: They can flip a coin to lose $200 for heads and nothing for tails, or they can skip the toss and pay $100 immediately. Most people will take the gamble. To the imagined rational man, the two games are mirror images; the choice to gamble or not should be the same in both. But to a real, irrational man, who feels differently about loss than gain, the two games are very different. The outcomes are different, and sublimely irrational.

2) Assumption: All investors are alike.

Theory:

People have the same investment goals and the same time-horizon; they all aim to measure their returns and fold their cards after the same holding period, whether days or years. Given the same information, they would make the same decisions. While their wealth may vary, none of them is rich or powerful enough to influence prices on their own. They have, in the terminology of economics, homogeneous expectations. They are price-takers, not makers. They are like the molecules in the perfect, idealized gas of a physicist: identical and individually negligible. An equation that describes one such investor can be recycled to describe all.

Reality:

Patently, people are not alike—even if differences in wealth are disregarded. Some buy and hold stocks for twenty years, for a pension fund; others flip stocks daily, speculating on the Internet. Some are "value" investors who look for stocks in good companies temporarily out of fashion; others are "growth" investors who try to catch a ride on rising rockets. Once you drop the assumption of homogeneity, new and complicated things happen in your mathematical models of the market. For instance, assume just two types of investors, instead of one: fundamentalists who believe that each stock or currency has its own, intrinsic value and will eventually sell for that value, and chartists who ignore the fundamentals and only watch the price trends so they can jump on and off bandwagons. In computer simulations by economists Paul De Grauwe and Marianna Grimaldi at the Catholic University of Leuven, in Belgium, the two groups start interacting in unexpected ways, and price bubbles and crashes arise, spontaneously. The market switches from a well-behaved "linear" system in which one factor adds predictably to the next, to a chaotic "non-linear" system in which factors interact and yield the unanticipated. And that is with just two classes of investors. How much more complicated and volatile is the real market, with almost as many classes as individuals?

3) Assumption: Price change is practically continuous.

Theory:

Stock quotes or exchange rates do not jump up or down by several points at a time; they move smoothly from one value to the next. Continuity of this sort characterizes all physical systems subjected to inertia; it is, for instance, the way temperature rises and falls during the day. And it jumped long ago into economics theory: *Natura non facit saltum* or, Nature does not make leaps, was the motto of one of the discipline's first reference texts, the 1890 *Principles of Economics* by Alfred Marshall. If you assume continuity, you can open the well-stocked mathematical toolkit of continuous functions and differential equations, the saws and hammers of engineering and physics for the past two centuries (and the foreseeable future). You can also draw important, useful inferences. For instance, as discussed in the preceding chapter, Markowitz's central idea was to reduce all investment decisions to two simple numbers, the mean and variance of expected prices, mathematical proxies for return and risk. In 1970 MIT's Samuelson offered a proof for Markowitz's simplification predicated on the assumption that prices change continuously.

Reality:

Clearly, prices do jump, both trivially and significantly. The trivial: Brokers often quote prices in round numbers, skipping intermediate values. Thus in the currency market, professional traders observe, about 80 percent of quotes end in a 0 or a 5, skipping the intermediate digits. The usual odds would suggest those values, being just two of the ten possible final digits in a number, should occur only about 20 percent of the time. Then there is the significant: Almost every day on the New York Stock Exchange, "order imbalances" occur in one stock or another. On one typical day, January 8, 2004, Reuters News Service reported imbalances happening eight times. Here, major news—approval of a medicine by the Food and Drug Administration, an unexpected takeover offer, or a windfall legal victory—caused market indigestion; sell and buy orders did not match, and market-makers had to raise or lower their price quotes until they did. To cope, some exchanges license "specialist" broker-dealers to step into the breach and trade when others will not. These specialists, while risking much, also profit greatly. Discontinuity, far from being an anomaly best ignored, is an essential ingredient of markets that helps set finance apart from the natural sciences.

4) Assumption: Price changes follow a Brownian motion.

Theory:

Brownian motion, again, is a term borrowed from physics for the motion of a molecule in a uniformly warm medium. Bachelier had suggested that this process can also describe price variation. Several critical assumptions come together in this idea.

First, *independence*: Each change in price—whether a five-cent uptick or a $26 collapse—appears independently from the last, and price changes last week or last year do not influence those today. That means any information that could be used to predict tomorrow's price is contained in today's price, so there is no need to study the historical charts.

A second assumption: statistical *stationarity* of the price changes. That means the process generating price changes, whatever it may be, stays the same over time. If you assume coin tosses decide prices, the coin does not get switched or weighted in the middle of the game. All that changes are the number of heads or tails as the coin is tossed; not the coin itself.

And a third assumption: the *normal* distribution. Price changes follow the proportions of the bell curve—most changes are small, an extremely few are large, in predictable and rapidly declining frequency.

Reality:

Life is more complex. This third set of assumptions is the one most clearly contradicted by the facts. Because it underpins almost every tool of modern finance, it gets special attention in the following chapter-in-a-chapter.

Pictorial Essay:
Images of the Abnormal

PICTURES ARE UNDERVALUED in science. They are not trusted. That is partly the 200-year-old legacy of the French mathematicians Lagrange and Laplace, who scrupulously labored to reduce all logical thought to precise formulae and carefully chosen words; sloppy diagrams were suspect. Their motivation was, I believe, partly technological: At that time drawings were imprecise and costly, a product of human hands. But in our lifetime the computer has changed all that. A modern diagram or chart can be as precise as desired, and is no more costly than the computer that draws it. The picture can now aid, not mislead (or replace!) the scientist. It permits instant comparison, instant comprehension. Thus we begin this assault on the normal with pictures, not numbers.

Start by looking more closely at real price charts. They are so common in newspapers and on television that, by their familiarity, their intricacy can be easily missed. First, consider the most frequently published chart of all, the Dow Jones Industrial Average. It is a simple average of the stock prices of the thirty most-highly valued companies in the United States. There are scores of other indices, with more or fewer stocks, varying criteria for inclusion, and different weighting systems. But the Dow, due to its age, simplicity, and wide following, is a good place to look first. It is the Mona Lisa of pictures in financial markets. So we examine it in this pictorial essay. In successive steps, we clean the years of accumulated grime from its surface to show the real information it conveys. And so we come to understand its enigmatic smile.

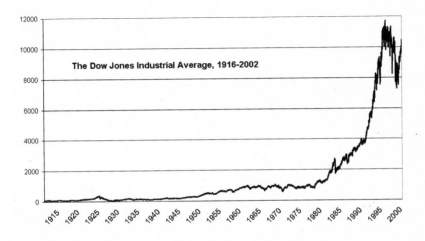

The Dow Jones Industrial Average, 1916-2002

The Old Master. Here is the Dow in its most familiar form: The actual, daily index values, from 1916 through to its peak of 11,722 in January 2000 and the few years of bear market that followed.

Prominent features: Few, aside from the broad upward trend. The spike downwards, of October 19, 1987, is visible. But what stands out is the rocket rise of the 1990s. For the most part, this way of drawing the Dow makes it appear as if history did not begin until about the 1980s, when the index finally left the 1,000-mark behind.

Try another approach, to see more.

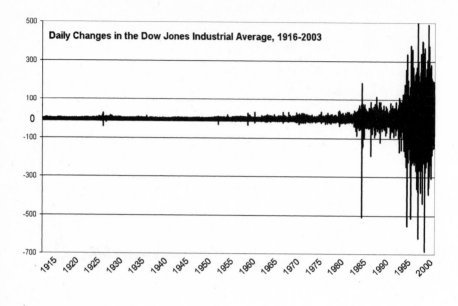

Looking closer. Here is the same data—but rather than showing the index values, this displays the index changes from one day to the next.

Prominent features: The magnitude of the index fluctuations increased towards the end of the twentieth century, as you would expect with a rapidly rising index value: Whereas in 1900 a 1 percent increase was a one-point rise in the Dow, by 2000 it was a 100-point rise. But you can see that, even in a market that rises overall, you can still get many vertiginous, one-day falls.

Now, draw the index a different way, to see more.

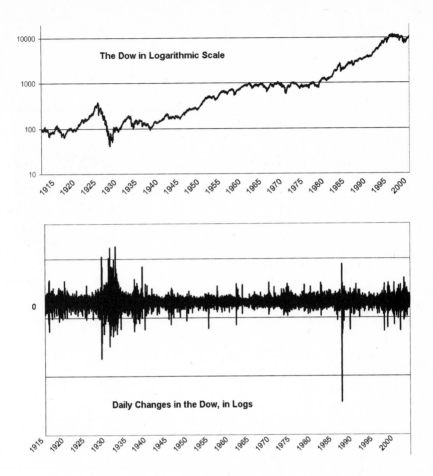

Looking under the varnish. Two charts here: the same daily index values (top chart) and changes (bottom chart) as before—but drawn to a more useful scale, the logarithmic. Logarithms rescale everything, so that a 1 percent change in 1900 will look about the same on our charts as a 1 percent change in 2000. That is just a different way of looking at the data. It makes the charts look the way the market actually felt to someone living through it.

Prominent features: The overall change in the magnitude of the index is no longer overwhelming. The Crash of 1929, the Great Depression, and World War II dominate the picture—just as they dominate our understanding of twentieth-century American economic history. Only the Crash of 1987 rivals those turbulent years. But most price changes merge into a broad strip, which varies in some sort of irregular pattern. The strip alternately narrows and widens, in some apparently haphazard cycle of thin and broad. Also, the spikes seem most likely to cluster together when the strip is wide.

Now we put the Dow to one side, and look at some new data.

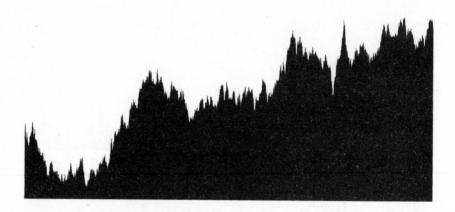

The reproduction. These two charts use the same drawing methods we applied to the Dow—but the picture is very different. These are price charts according to the Bachelier Brownian motion model. As discussed earlier, this is in the catechism of orthodox financial theory. It assumes each day's price change is independent of the last, and follows the mildly random pattern predicted by the bell curve. The top chart shows a computer-simulated Brownian price series—the silicon version of the previous, ordinary Dow chart. The bottom charts the changes from one Brownian moment to the next.

Prominent features: By "eyeball" comparison with the Dow, this is not merely different from it; it is an entirely distinct species. While the topmost chart could pass for reality, the bottommost chart is obviously aberrant. Compared to the Dow, this chart's spikes rise and fall within a small range like the blades of grass in a lawn. Its tallest spikes are interspersed across the entire chart, rather than concentrated into moments of high drama. Let us magnify the contrast, with a new scale.

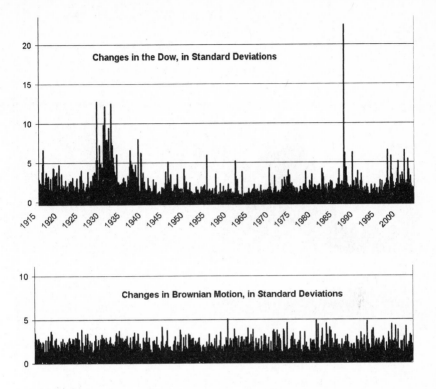

Original vs. reproduction—through the analyzer. Here you can see the differences between the Brownian (bottom) and Dow (top) charts more clearly. Instead of using a log scale as before, here we translate each index change into the number of standard deviations it is beyond the average change—in other words, how unusual it is. A very large, rare index movement will have a tall bar on this chart; the common, small changes have short bars.

Prominent features: In the Brownian chart, most changes—in fact, about 68 percent—are small. They are within one standard deviation of the average index change, zero. Mathematicians use the Greek letter sigma, σ, for standard deviation. About 95 percent of the changes are within 2σ, 98 percent within 3σ, and very, very few values are any larger. Next look at the Dow variations. The spikes are huge. Some are 10σ; one, in 1987, is 22σ. The odds of that are something less than one in 10^{50}—so minute that the standard Gaussian tables do not even contemplate it. In other words, virtually impossible. Yet there it is.

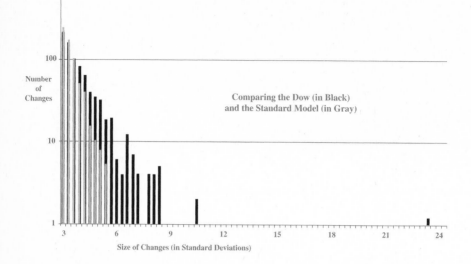

Two into one will not go. Here, the preceding two charts are super-imposed on each other and mounted in a new frame. We take the daily differences for each index and reshuffle them so that they are ranked by size, rather than by the order in which they happened. Increments of the same size are counted, and the result is plotted. The very big changes, plus or minus, are shown at the right end of the chart; the very small changes, plus and minus, are clustered at the extreme left.

Prominent features: The Brownian data (gray bars) tail off rapidly. This particular simulation—it being a random game, precise results will differ each time you play it—has no changes greater than about 5σ. The Dow data are in black and continue under the gray bars. They spill out beyond the narrow confines of the Brownian model. It has many changes beyond 5σ—and one at 22σ. This is the "fat tail" to which stat-isticians refer. And it means the standard model of finance is wrong.

The Evidence

Pictures of one stock index are instructive, but should not be considered alone. We should also check other data, other indices, other markets, using other tools. And that is exactly what many economists have been doing for the past four decades. Herewith, a flavor of the results.

Commodities

Years after Bachelier's thesis, a few other researchers began checking the data themselves and observed some disturbing trends. But they largely discounted these dissonant data as aberrations to be ignored. As mentioned earlier, the origins of Gaussian analysis in astronomy conditioned scientists to assume that, in this messy real world, there would always be a few anomalous bits of information, outliers, that experimental error or a capricious nature would provide to spoil the scientist's tidy pictures. Generally, these extreme values are simply discarded as errors, ignored before the main data-crunching begins.

Thus it was not until 1962 that the first substantial body of contradictory data appeared. I analyzed more than a century of data on U.S. cotton prices and studied the way they had varied daily, monthly, and yearly. The results were clear and irrefutable. Far from being well-behaved and normal as the standard theory then predicted, cotton prices jumped wildly around. Their variance, rather than holding steady as expected, gyrated a hundred-fold and never settled down to a constant value. In the world of financial theory, that was a bombshell. When Cootner of MIT reprinted my analysis in his book a year later, he wrote that it forced economists "to face up in a substantive way to those uncomfortable empirical observations that there is little doubt most of us have had to sweep under the carpet up to now." The paper, one of the most widely read and cited in economics, sparked others to look at the price data with fresh eyes. Because of its import, I will come back to this tale.

Stocks

The inquiry quickly broadened beyond cotton. Whatever the stock index, whatever the country, whatever the security, prices only rarely follow the predicted normal pattern. My student, Eugene Fama, investigated this for his doctoral thesis. Rather than examine a broad market index, he looked one-by-one at the thirty blue-chip stocks in the Dow. He found the same, disturbing pattern: Big price changes were far more common than the standard model allowed. Large changes, of more than five standard deviations from the average, happened two thousand times more often than expected. Under Gaussian rules, you should have encountered such drama only once every seven thousand years; in fact, the data showed, it happened once every three or four years.

Later researchers have found much the same thing in stock indices. Statisticians like to condense a lot of confusing information into one clear talking point, and so they have devised a single number to measure what we have been discussing—how closely real data fit the ideal bell curve. They call it kurtosis, from the Greek *kyrtos*, or curved. But we can think of it as how much "spice" is in the statistical broth. A perfect, unseasoned bell curve has a kurtosis of three. A hot, fat-tailed curve of the sort we have been finding would have a higher spice number, while a curve that had been boiled into a dull paste would have a lower number. According to a 2003 book by Wim Schoutens, a Catholic University of Leuven mathematician, the daily variations in another common U.S. stock-market index, the Standard & Poor's 500, had a kurtosis of 43.36 between 1970 and 2001. This is, by the bland standards of the statistical kitchen, a five-alarm chili. If you throw out the spiciest data point, the October 1987 crash, you still get an uncomfortably hot dish: a kurtosis of 7.17. The high-tech Nasdaq index: 5.78. The French CAC-40: 4.63. All are above the Gaussian norm of three.

· · ·

Currencies

Evidence abounds of abnormal foreign-exchange markets. A Citigroup study in 2002 found unpleasantly sharp price swings in several currencies—dollar, euro, yen, pound, peso, zloty, even the Brazilian real. On one day, the dollar vaulted over the yen by 3.78 percent. That is 5.1 standard deviations, or 5.1σ, from the average. If exchange rates were Gaussian that would be expected to happen once in a century. But the biggest fall was a heart-stopping 7.92 percent, or 10.7σ. The normal odds of that: Not if Citigroup had been trading dollars and yen every day since the Big Bang 15 billion years ago should it have happened, not once.

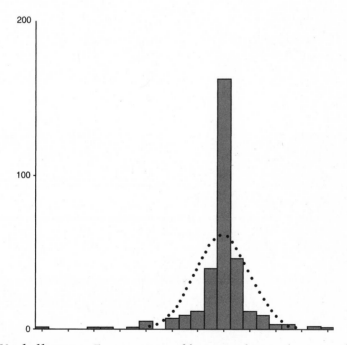

No *bell* curve. Four centuries of history and turmoil are recorded here, in this record (from DeVries 2002) of the frequency of different-size changes in the sterling-guilder exchange rate. These data, from 1609 to 2000, do not fit the standard bell curve: There are too many price changes that are very small, and too many that are very large—hence too few points in between.

The same phenomenon was found in daily, weekly, monthly, and annual exchange rates, though the kurtosis, or abnormality, diminishes as the time stretches out. Yet this is a not a new phenomenon: The same picture, of real price data not fitting within the bell curve, emerges from records of British pound-Dutch guilder exchange going all the way back to 1609.

Dependence

Of course, well-behaved price changes are not the only assumption underlying the standard financial model. Another is that each flip of the coin, each quiver of the price, should be independent of the last. There should be no predictable pattern on which you could trade and profit. Alas for the financial establishment, this is also a fairy tale.

The most-studied evidence, by the greatest number of economists, concerns what is called short-term dependence. This refers to the way price levels or price changes at one moment can influence those shortly afterwards—an hour, a day, or a few years, depending on what you consider "short." A "momentum" effect is at work, some economists theorize: Once a stock price starts climbing, the odds are slightly in favor of it continuing to climb for a while longer. For instance, in 1991 Campbell Harvey of Duke—he of the CFO study mentioned earlier—studied stock exchanges in sixteen of the world's largest economies. He found that if an index fell in one month, it had slightly greater odds of falling again in the next month, or, if it had risen, greater odds of continuing to rise. Indeed, the data show, the sharper the move in the first month, the more likely it is that the price trend will continue into the next month, although at a slower rate. Several other studies have found similar short-term trending in stock prices. When major news about a company hits the wires, the stock will react promptly—but it may keep on moving for the next few days as the news spreads, analysts study it, and more investors start to act upon it.

Just the opposite appears to happen in the medium term, three to eight years. A stock that was rising over one multi-year stretch has slightly greater odds of falling in the next. A 1988 study by Fama and another economist, Kenneth R. French, documented this. They looked back over the price records of hundreds of stocks and grouped them into portfolios based on their size. They found that about 10 percent of a stock's performance in one eight-year period could be attributed to how it did in the prior eight-year period—that is, there was a small but measurable tendency for a stock doing well in one decade to do poorly in the next. The effect was weaker, but still statistically significant, at shorter time-scales of three to five years. Others have corroborated such findings.

A "fad" effect may be happening, some economists theorize. For a few years, a company can be in favor among investors: Its products are selling, its earnings rising, and its investors projecting even better times to come. Then something happens to break the mood: The company stumbles, or investor fashions change. The price trend reverses. A "correction" sets in. The effect is not great; but some economists think it could be, at times, large enough to make money from—a flat violation of the standard theories. In 1993 two economists, Narashimhan Jegadeesh and Sheridan Titman, constructed an elaborate test of trading strategies based on these trends. Using market data from 1965 to 1989, they simulated what would have happened if they had followed a simplistic strategy: Buy stocks that had risen in the prior six months, and sell those that had fallen. They found they could have made a tidy paper profit in the following six-month period, on average, 12.01 percent a year above what a simple, market-following index fund would have earned them. But beyond six months, the picture changed: After two years, their paper profits vanished as the stock prices "corrected" themselves.

Their results, as that of similar studies, are controversial. Critics claim they overestimate the profits and underestimate the costs of such a trading strategy. I, myself, mistrust them for other reasons: When a statistician finds a result he had been expecting, he tends not to put his tests under as critical a microscope as he should—especially when he is also assuming a Gaussian world. As will be seen in a later chapter, I have another view of dependence. To me, its most important effect is not over a short term, but over the very long term—in theory, an infinite effect. This has some unusual consequences. Be that as it may, for our present purposes, a bottom line emerges: Stock prices are not independent. Today's action can, at least slightly, affect tomorrow's action. The standard model is, again, wrong.

But Does It Work?

Such is the weight of evidence against the assumptions in the standard model that it is no longer reasonable to ignore   entirely. Indeed, forty years after I started a battle on the subject, most economists now acknowledge that prices do not follow the bell curve, and do not move independently. But for many, after acknowledging those points, their next comment is: So what? Independence and normality are, they argue, just assumptions that help simplify the math of modern financial theory. What matters are the results. Do the standard models correctly predict how the market behaves overall? Can an investor use Modern Portfolio Theory to build a safe, profitable investment strategy? Will the Capital Asset Pricing Model help a financial analyst, or a corporate finance officer, make the right decision? If so, then stop arguing about it. This is the so-called positivist argument, first advanced by University of Chicago economist Milton Friedman.

Alas, by that measure, too, the standard tools of financial theory often fail. Economics is a faddish discipline. In the 1970s, when the CAPM and Black-Scholes ideas were spreading, the way to get ahead in economics was to find evidence that they were right. So, evidence was found and dissent ignored. But in the 1980s, a correction set in that has continued to this day. Young economists see Sharpe and his generation as old boys, to be challenged. Bit by bit, new evidence has been emerging in academic journals and Wall

Street newsletters that reality is more complicated than the old-style religion allows.

Recall that, under CAPM, the return an investor should expect to receive from a stock is just the T-bill rate, plus some proportion of the stock-market's overall performance; that proportion is the crucial "beta" value, which varies from stock to stock. Under the orthodox theory, nothing else should be going on. No need to study the fundamentals of the company in question. No need to pump friends on the company's board for inside information. Just calculate the beta, check the T-bill rate in the newspaper, and make a broad economic forecast about how the stock market overall will do. End of story.

In fact, the story is a lot longer than that. A string of what economists euphemistically call "anomalies" have been found—effects that do not fit or that contradict CAPM:

Anomaly 1: *The P/E Effect*. Financial analysts often compare a stock price to other numbers to help decide whether it is expensive or cheap. The most common tool is the price/earnings ratio: the stock price divided by the company's per-share earnings. Orthodox theory calls that a waste of time: Only beta, the degree to which a stock does or does not move with the rest of the market, should matter to its price. P/E should be meaningless. In fact, several studies have found, stocks with high P/E ratios tend to perform worse than stocks with low ratios. That is, of course, just common sense: A stock for which you overpay from the start is less likely to give you a profit.

Anomaly 2: *The Small-Firm-in-January Effect*. Shortly after the P/E factor was studied, economists discovered the "January effect" mentioned earlier: a clear tendency of the market to rally every January. Then, a "small-firm effect" was discovered: Portfolios of small-company stocks outperformed large companies by 4.3 percent, economists found. And, further study found, a "small-firm-in-

January" effect combining the two phenomena was even more pronounced than either on its own. Again, the orthodox financial theory wishes these effects away. When a statistician looks for
correlations between prices and various factors that could be affecting them, only the stock-market beta should pop out as having any
importance whatsoever.

Anomaly 3: *The Market-to-Book Effect.* Another common
financial ratio used by stock-pickers is market-to-book: That is,
divide the stock price by the per-share value that the company's
accountants report in the financial reports, or "book." Surprise:
Companies with low ratios—that is, those that the stock market values less than does the company's accountant—perform better over
time than companies with high ratios. Of course, this is nothing
more than the old Wall Street mantra, buy low, sell high. And
again, by the standard theories, it should not work.

Many more such anomalies have been reported in economics
journals. But this kind of research came to fruition in an especially
influential 1992 paper by Fama and French. They tried to create the
economic equivalent of a double-blind drug trial, devising tests and
controls to prevent any unintended bias from slipping into the
results. They looked at the price/earnings effect and the
market/book effect—and found those two factors alone could
account for most of what differentiated the profitability of one stock
from another. Beta was redundant. It was, Fama and French
asserted, "a shot straight at the heart of the (CAPM) model." That
phrase has earned their work a shorthand title among other economists: the beta-is-dead paper.

So much for CAPM. As for Black-Scholes, the original options-
pricing formula is now widely accepted to be imprecise at best, and
misleading at worst. Finally, an especially lively pastime for economists these days is to try poking holes in the grand unified theory of
modern finance, the Efficient Markets Hypothesis that markets are

rational, prices reflect all available information, and you cannot beat the market. In fact, it appears, sometimes you can. By 1989, Peter Lynch, one of the most successful investment managers, had guided Fidelity's Magellan Fund to beat the market index in eleven out of thirteen years. The odds of Lynch accomplishing that by dumb luck, as the Bachelier model would have it, are slim but not impossible: About one chance out of 105, according to one study. But it was not just the frequency of success that was striking about Magellan; its magnitude was more unusual. The fund's average annual return for the entire period was 28 percent, compared to 17.5 percent for the Standard & Poor's 500 index. And for its first seven years— when it was still a small fund, too small for any detractors to argue that its size alone gave it a competitive edge in the marketplace— Magellan beat the market by an average 25 percent a year. The odds of that occurring by dumb luck are less than one in 10,000—"far beyond the bounds of luck in an efficient market," concluded the study's author, Alan J. Marcus, a Boston College finance professor.

The Persistence of Error

Then why, with so much evidence against the orthodox financial models, do most economists still teach them, and why do many financiers honor them? If this were astronomy, the argument would have ended long ago. Imagine observatories suddenly finding a new planet where, the standard theory says, none should be. And then another, and another and another. Astronomers, after checking their instruments, would not ignore the data; they would question their understanding of celestial mechanics and a new and fruitful episode in astronomy would dawn. But it does not work that way in economics, even though the equivalent of countless new planetary sightings have been recorded. In part, the profession's reaction reflects the nature of finance and statistics; there are few open-and-shut cases when an economist meets a computer database. Yes, some of the individual arguments against the standard model are by now irrefutable: Prices are, indeed, abnormal and dependent. Some other arguments, such as "beta is dead," are strong but not bullet-proof; millions of words have gone, in the academic press, to critique Fama and French's paper.

And the high priests of modern financial theory keep moving the target. As each anomaly is reported, a "fix" is made to accommodate it. When CAPM first came under attack, academic economists devised a broader model, called Arbitrage Pricing Theory. Rather than work with just one factor, beta, APT incorporates as many factors as desired: a beta for the market/book effect, a beta for the price/earnings effect, a beta for the state of the economy, and a beta for any other factor that could conceivably affect stock prices. Likewise, when it became clear that volatility really does cluster and vary over time rather than stay fixed as the standard model expects, economists devised some new mathematical tools to tweak the model. Those tools, part of a statistical family called GARCH (a name only a statistician could love), are now widely used in currency and options markets.

But such ad hoc fixes are medieval. They work around, rather than build from and explain, the contradictory evidence. They are akin to the countless adjustments that defenders of the old Ptolemaic cosmology made to accommodate pesky new astronomical observations. Repeatedly, the defenders added new features to their ancient model. They began with planetary "cycles," then corrected for the cycles' inadequacies by adding "epicycles." When these proved inadequate, yet another fix moved the center of the cycles away from the center of the system. In the end, they could fit all of the anomalous data well enough. As more data arrived, new fixes could have been added to "improve" the theory. They satisfied their early customers, astrologers. But could they lead to space flight? It took the combined efforts of Brahe, Copernicus, Galileo, and Kepler to devise a simpler model, of a sun-centered system with elliptical planetary orbits. The long and well-documented history of successful sciences includes many such examples of pyramids of fixes—but they are viewed as stopgaps.

So again, why does the old order continue? Habit and convenience. The math is, at bottom, easy and can be made to look impressive, inscrutable to all but the rocket scientist. Business schools around the world keep teaching it. They have trained thousands of financial officers, thousands of investment advisers. In fact, as most of these graduates learn from subsequent experience, it does not work as advertised; and they develop myriad ad hoc improvements, adjustments, and accommodations to get their jobs done. But still, it gives a comforting impression of precision and competence.

It is false confidence, of course. The problem lies at the roots of the standard model, in its assumption that the best way to think about stock markets is as a grand game of coin-tossing. If you are going to use probability to model a financial market, then you had better use the right kind of probability. Real markets are wild. Their price fluctuations can be hair-raising—far greater and more damaging than the mild variations of orthodox finance. That means that individual stocks and currencies are riskier than normally assumed. It means that stock portfolios are being put together incorrectly; far from managing risk, they may be magnifying it. It means that some trading strategies are misguided, and options mis-priced. Anywhere the bell-curve assumption enters the financial calculations, an error can come out.

History is replete with ironies. And it is one of the greatest that the truly wild nature of markets was re-discovered, at their cost, by two of the most ardent formulators of orthodox economics, Scholes and Merton. In 1993, the two Nobel laureates joined some heavyweight Wall Street bond traders in the creation of a new hedge fund, Long-Term Capital Management LP. The partners collectively contributed $100 million and raised a war-chest that eventually topped $7 billion. Their strategy was straightforward. They would scour the world for occasions when, by their orthodox valuation formulae, the prices of individual options appeared to be wrong. They would bet heavily—with a "leverage" or debt ratio as great as 50-to-1—on the market's eventually correcting the mistake. They had at one point twenty-five Ph.D.'s on the payroll. As Sharpe, an onlooker to the fund, told the *Wall Street Journal,* LTCM "was probably the best academic finance department in the world."

But it blew up. After profits of 42.8 percent in 1995 and 40.8 percent in 1996, the fund in 1998 hit turbulent markets. It had already started straying from the pure academic strategy, taking hyper-risky bets on the direction of bond prices rather than just on market "mistakes"—much to the dismay of Scholes. Then world tensions began mounting, and bond prices began doing things that the models had not forecast. The fund started losing money. In August 1998 the Russian government defaulted on its bonds, triggering a market meltdown. LTCM had been one of the biggest Western traders in the bonds, and was stuck without buyers. Worse, contrary to the academic predictions, most of the fund's other investments started failing, too. Global markets, far from displaying independent price

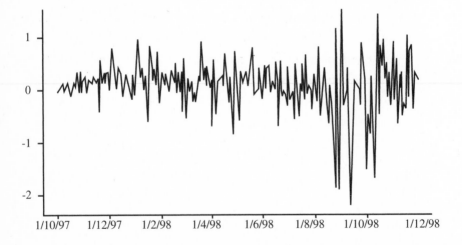

A wild market. At the height of Russia's 1998 debt default, global markets passed through a hurricane. This diagram, from Medova 2000, shows the aggregate, daily profits and losses of four of the world's biggest banks during that period, as they tried to cope with bucking foreign exchange markets.

changes, were suddenly marching all in the same direction at once: down. The same with volatility: up. After the fact, some economists studied the market, and the record they compiled of the market's manic state is truly impressive. They looked at the daily profits and losses of four of the biggest global banks, as they invested in the currency markets. To preserve the banks' anonymity, they aggregated the data into one series. But the effect is striking, nonetheless. The swings—up and down—at the height of the crisis show just how wild markets can be.

In the end, several banks reluctantly agreed to bail out the fund through a $3.625 billion takeover. That came only at the behest of the Federal Reserve Board, which was concerned about a wave of bankruptcies if LTCM went under. Scholes himself later denied that the option-pricing models played any but "a minor role" in the debacle. But some of his partners do not see it quite that way. John Meriwether, the fund's prime mover and the man who may have lost the most, $150 million, told the *Wall Street Journal*: "Our whole approach was fundamentally flawed." In launching a new fund in 2000 (Wall Street folk are nothing if not resilient), he observed: "With globalization increasing, you'll see more crises. Our whole focus is on the extremes now—what's the worst that can happen to you in any situation—because we never want to go through that again."

Amen.

PART TWO

· · · ·

The New Way

The classical theorists resemble Euclidean geometers in a non-Euclidean world who, discovering that in experience straight lines apparently parallel often meet, rebuke the lines for not keeping straight—as the only remedy for the unfortunate collisions which are occurring. Yet, in truth, there is no remedy except to throw over the axiom of parallels and to work out a non-Euclidean geometry. Something similar is required today in economics.

—John Maynard Keynes

CHAPTER VI

Turbulent Markets:
A Preview

SO WE COME TO THE CRUCIAL QUESTION: If the theorists were wrong about financial markets for so many years, then how to set ourselves straight? The answer that I propose comes from an unlikely quarter, blowing in the wind.

Wind is a classic example of a form of fluid flow called turbulence. Though studied for more than a century, turbulence remains only partly understood by either theoreticians or aircraft designers. Wire a wind tunnel at Boeing or Airbus with appropriate instruments; and you can detect the complex motion of the water vapor, dust, or luminescent markers blowing inside it. When the rotor at the tunnel's head spins slowly, the wind inside blows nice and smoothly. Its currents glide in unison in long, steady lines, planes and curves like parallel sheets of supple, laminated plywood. This kind of flow is called laminar. Then, as the rotor accelerates, the wind inside the tunnel picks up speed and energy. Here and there, it suddenly breaks into gusts—sharp, intermittent. This is the onset of turbulence. The wind inside the tunnel dissipates the rotor's energy.

Eddies form; and on those eddies yet more, smaller eddies form. A cascade of whirlpools, scaled from great to small, spontaneously appears. Then, just as suddenly, a surprise. Here and there, smooth flow returns momentarily. And then more gusts and turbulence. Smooth again. Rough again. In a real jet, flying high above the ground, you can feel this on-off turbulence in the bump, bump, bump of the craft buffeted every so often on gusting updrafts, downdrafts, and eddies. In a smaller plane, a more sensitive probe to the wind's whims, you can feel it with greater violence. Here, in an illustration from one of my papers in 1972, you can see this intermittent motion.

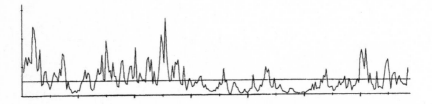

A turbulent wind: In the atmosphere. A chart of the original multifractal simulation (Mandelbrot 1972) of changing wind speed as it bursts into and out of gusty, turbulent flow. Notice how the peaks and troughs cluster together.

Now, this is old knowledge—so old that its significance can be overlooked. We see turbulence almost any day, just looking up at the billows upon billows of a cumulus cloud. We see it through a telescope, in Jupiter's celebrated red eye. We see it through a spectrometer, in the pattern of sunspots. In this age, thanks to airborne movie or television cameras, we can see it in a news report of a disabled oil tanker; its oily wake spreads behind it in an awesome but beautiful pattern of swirls and eddies. Telephone engineers, too, can hear the signature of turbulence—the intermittency, the energy fluctuations—in what they call electronic "flicker noise," the irregular and inexplicable pops and crackles that, despite the greatest precautions, cause errors in data transmission. Artists observed turbulence without waiting for scientists to guide them. Its power impressed Leonardo da Vinci:

> Amid all the causes of the destruction of human property, it seems to me that rivers hold the foremost place on account of their excessive and violent inundations. . .
>
> Against the irreparable inundation caused by swollen and proud rivers no resource of human foresight can avail; for in a succession of raging and seething waves gnawing and tearing away high banks, growing turbid with the earth from ploughed fields, destroying the houses therein and uprooting the tall trees, it carries these as its prey down to the sea which is its lair, bearing along with it men, trees, animals, houses, and lands, sweeping away every dike and every kind of barrier, bearing along the light things, and devastating and destroying those of

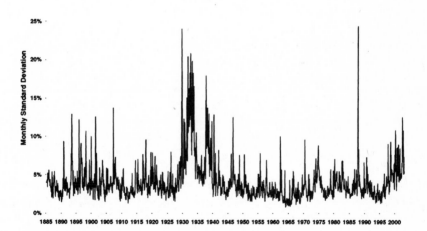

A turbulent wind: In the market. This chart, from Schwert 2004, shows the changing volatility of the stock market, as the magnitude of price changes varied wildly from month to month. Peak activity is 1929–1934, and again in 1987. The similarity to the wind chart shown earlier is uncanny—as you would expect when comparing data from two turbulent systems.

weight, creating big landslips out of small fissures, filling up with floods the low valleys, and rushing headlong with destructive and inexorable mass of waters.

From the Notebooks

That same kind of turbulence, often destructive, is visible in financial markets. In fact, the chart on page 113 illustrates this. It shows how the volatility of the stock market has been, itself, volatile—varying wildly through the turbulent twentieth century. If you compare this with the earlier wind chart, you can see the same bump, bump, bump; the same abrupt lurches between wild motion and quiet activity; the same discontinuities; the same intermittency; the same concentration of major events in time. Think about that small plane again, juddering as it crosses a turbulent air stream: Can that be analogous to a white-knuckled investor watching a stock price buck and bump beneath him?

Turbulent Trading

Certainly, turbulence is a common metaphor for financial commentators, and it is easy to see why. For a graphic example, look back at the New York Stock Exchange on October 27, 1997. That day saw the Dow Jones Industrial Average, or DJIA, lose a heart-stopping 554.26 points, or 7.18 percent. After the fact, as so often in these cases, the fatal trigger was hard to identify, though staffers of the U.S. Securities and Exchange Commission spent a year trying to reconstruct events. But the impact was profound. Cascades of selling washed across the exchange—forcing authorities to halt trading twice, in a vain effort to calm people down. Listen to the action, as summarized in the "just-the-facts" language of the final SEC report:

At 2:36 p.m. on October 27, the DJIA had declined a total of 350 points from the previous trading session's closing value. This decline triggered a 30-minute halt on the stock, options and index futures markets. After stock trading resumed at 3:06 p.m., prices fell rapidly to reach the 550-point circuit breaker level at 3:30 p.m., thereby ending trading 30 minutes prior to the normal stock market close.

On Tuesday, October 28, market prices initially resumed their sharp decline. By 10:06 a.m. the DJIA had declined a total of 187.86 points (2.62%). The market subsequently rallied sharply, with the DJIA closing up 337.17 points (4.71%) at 7498.32 on then-record share volumes of over a billion shares each on the NYSE and the Nasdaq Stock Market.

SEC Division of Market Regulation, September 1998

Truly a turbulent scene. It sounds like Leonardo's flood waters, bursting one dam after another before subsiding. As these financial waters raged, price quotations leaped wildly. The "spreads" between brokers' bid and ask prices widened sharply—as much as 19 percent above the industry's norms (that translates into an instantaneous windfall to any broker who called it right, and near-ruin to those who got it wrong). The turmoil spread around the globe: The Hong Kong index fell 14 percent, London 9 percent. In the final twenty-four minutes before the New York market closed at 3:30, prices plummeted at an average rate of 0.10 percent a minute, or 6 percent an hour, the SEC calculated. Put that into perspective: The value of American business was falling $100 million *a second*. The next morning, prices roared in the opposite direction even faster. But the fastest action of all concentrated into three isolated minutes in the whole twenty-four hours: between 3:12 and 3:14 p.m. New York time, and between 3:24 and 3:25 p.m. This was no mere financial storm. It was a hurricane.

Interesting, you say—but is this "turbulent markets" idea just a trope? Can you seriously compare the wind to a financial market, a gale to a rally, a hurricane to a crash?

In terms of the underlying causes, certainly not. But mathematically, yes. It is an extraordinary feature of science that the most diverse, seemingly unrelated, phenomena can be described with the same mathematical tools. The same quadratic equation with which the ancients drew right angles to build their temples can be used today by a banker to calculate the yield to maturity of a new, two-year bond. The same techniques of calculus developed by Newton and Leibniz two centuries ago to study the orbits of Mars and Mercury can be used today by a civil engineer to calculate the maximum stress on a new bridge, or the volume of water to pass beneath it. Now, none of this means that the bridge, river, and planets work the same way; or that an archaeologist at the Acropolis should help price an Accenture bond. Likewise, the wind and the markets are quite distinct; one is a phenomenon of nature, the other a creature of man. But the variety of natural phenomena is boundless while, despite all appearances to the contrary, the number of really distinct mathematical concepts and tools at our disposal is surprisingly small. When a man goes to clear a jungle he has relatively few types of tools: To cut, perhaps a machete; to knock down, a bulldozer; to burn, fire. Science is like that. When we explore the vast realm of natural and human behavior, we find our most useful tools of measurement and calculation are based on surprisingly few basic ideas. When a man has a hammer, all he sees around him are nails to hit. So it should be no great surprise that, with our small number of effective mathematical tools, we can find analogies between a wind tunnel and a Reuters screen.

My life's work has been to develop a new mathematical tool to add to man's small survival kit. I call it fractal and multifractal geometry. It is the study of roughness, of the irregular and jagged. I coined its name in 1975. Fractal is from *fractus*, past participle of *frangere*, to break, as I was reminded by one of my sons' Latin dictionaries. The same root survives in many common words, including *fraction* and *fragment*. I developed these ideas over many decades of intellectual wanderings—pulling together many stray, forgotten, under-explored, and seemingly unrelated artifacts and issues of the mathematical past, extending them in every direction, and creating a new, coherent body of mathematics. Fractal geometry has come to be viewed as "natural." It is used today for an improbably diverse set of tasks: compressing digital images over the Internet, measuring metal fractures, analyzing brain waves in an EEG machine, designing ultra-small radio antennae, making better optical cables, and studying the anatomy of lung bronchia.

The methods of fractal geometry have become part of the toolkit of fluid dynamics, hydrology, and meteorology. Its power comes from its unique ability to express a great deal of complicated, irregular data in a few simple formulae. This power is especially clear in the case of multifractality, which is fundamental in the study of turbulence and also handy in financial markets. So I and others have, over the past few decades, been using fractal notions to study and build models of how markets work. This is a work in progress, indeed, one would have to say, a work barely begun despite forty years of effort. Subsequent chapters elaborate on fractals and their application to finance. But for now, I offer a small preview of what fractal geometry—even in its simplest, cartoonish renderings—can suggest.

Looney 'Toons for Brown-Bachelier

Economists love models. To assemble a few easily controlled inputs into a lifelike model is to understand something fundamental about the way the world works. Here, I take Bachelier's model and present a hint of a version that is even simpler and easier—so simple, in fact, that I hesitate to call it a model at all. To avoid misunderstanding, let us call it a cartoon. I use the term in the sense of the Renaissance fresco painters and tapestry designers: a preliminary sketch in which the artist tries out a few ideas, and which if successful becomes a pattern for the full *oeuvre* to come. It will give a flavor of what is possible with just a few fractal tools.

A fractal, again, is a pattern or shape whose parts echo the whole. If you look closely at the frond of a fern, for instance, you see it is made up of smaller fronds that, in turn, consist of even-smaller leaf clusters. Of course, you can run such thinking forwards as well as backwards; you can analyze the fern down into its smaller parts, as well as synthesize the fern up from the smaller parts. Start with the smallest leaf shoots as the fern unfolds from its bud; then watch as each shoot grows and generates more shoots, which in turn grow and generate yet more shoots until the fern is fully formed. Such is Nature's method. Financial fractals can copy the same trick: analyze, as well as synthesize, a stock chart. Below, I synthesize.

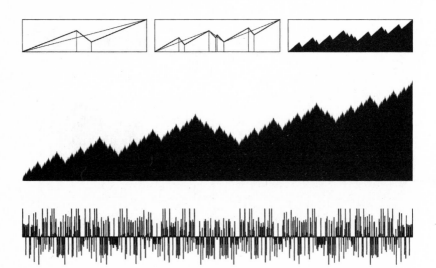

The cartoon stock chart. This shows how to construct a non-random cartoon of a fractal financial chart according to Bachelier. The top line, from left to right, shows the first stages. The central, black diagram is the completed fractal chart. The bottom shows the increments, from one moment to the next. This construction is of the simplest kind—not realistic, yet.

In the set of diagrams preceding, you see our financial fractal begins with a box, one unit wide by one unit tall (in our diagram, the width scale is stretched; but that is just to make a prettier picture). Inside the box, we draw a straight line rising from the bottom left corner, at coordinate (0,0), to the top right corner, at coordinate (1,1). This is the underlying trend line—the assurance that our final chart will eventually show a profit, no matter how much prices fluctuate along the way. If we wanted to model a market drop, we could as easily do so by starting with a line that falls

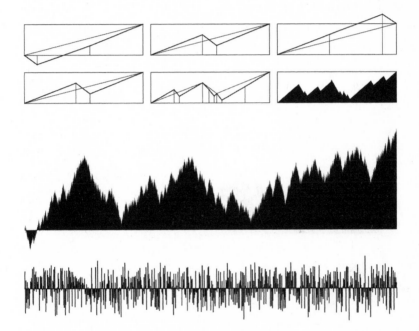

The randomized cartoon stock chart. This is similar to the chart on the facing page, but with pieces of its generator scrambled to add realism. As before, we start with the simple fractal generator, rising up, then falling down, and then rising again (shown here in the middle of the top row). Then we can shuffle the generators' pieces, into a down-up-up sequence (top left) or an up, up, down order (top right). At each stage of the fractal construction, we select one of the three possible generators at random. The second line shows the first three stages of the construction. The black "fever" chart is the completed diagram. The bottom line shows the changes.

from top left to bottom right. Then, you see a zigzag shape called generator that fits over the straight line. It is in three parts, as shown: It rises, breaks downward at a critical point, then breaks upward again. Exactly where it breaks and how frequently is crucial to the outcome.

Now come the instructions for building the fractal model. Wherever you see a straight line segment, replace it with a copy of the broken line that has been shrunken without being rotated. To make it fit, shrink it more horizontally than vertically. And to fit the endpoints of a down interval, flip the jagged shape over. Repeat, at each stage knocking out ever-smaller lines, and peopling the chart with ever-smaller zigzags. At each iteration, the curve becomes more irregular, more jagged. If you squint at the final box you can imagine a kind of price chart taking shape—but not very realistic, and far too predictable.

So far, the intervals of the zigzag generator point up, then down, and finally up again. To improve the realism, instead of blindly repeating this pattern, scramble the pieces. Before each step of the construction, roll the dice to pick a new order for the three segments of the generator: (up, up, down) or (down, up, up) or the original (up, down, up). Continue the process, and the new chart starts to look real.

Preview of More Close-Fitting Cartoons

The cartoon process—and its outcome—can be further complicated, as will be done in a later chapter. Change where the zigzag line breaks, or how often it breaks. Remove it from the rectangular box and let it grow more spontaneously. An infinite range of possibilities arises—and this, from a game with a straight line. For more complication, start working with data sets rather than lines on paper. Try statistical relations or other abstract patterns. Soon, the charts take on a startling realism. They can show the extreme, wild variations described earlier. They can show dependence, with patterns at the beginning of the chart influencing patterns at the end. As will be seen, they can replicate any kind of financial chart in full fidelity.

Of course, real price charts do not arise this way. Real charts merely record thousands of individual transactions, as they happened. But as I have said before, we cannot possibly analyze every motive behind every one of those transactions; such "fundamental" analysis is futile. Instead, what we can do is create a mathematical model that can mimic the real thing—can mimic how much a price varies, how quickly it rises or falls. It will not trace the identical, real path of the real price, but it will "behave" statistically in the same way. And from it, you can develop a powerful new tool to study and work in the market. You can compare the riskiness of one investment against another. You can play out, on a computer, "what-if?" scenarios with your portfolio. You can estimate the value of a complicated new financial product, an "exotic" option. And you can track—and perhaps forecast—how turbulent the market is becoming.

Suddenly, turbulence ceases to be a metaphor. Multifractals make turbulence a fundamentally new way of analyzing finance. Markets no longer appear in the entirely rational, well-behaved patterns of past financial theorists. They are seen for what they are: dynamic, unpredictable, and sometimes dangerous systems for transferring wealth and power, systems as important for us to understand as the wind, the rain, and the flood. And floods—natural or manmade—need defenses. Machiavelli once saw fortune as a flood, and his metaphor is apt here.

> I liken her [Fortune] to one of these violent rivers which, when they become enraged, flood the plains, ruin the trees and the buildings, lift earth from this part, drop in another; each person flees before them, everyone yields to their impetus without being able to hinder them in any regard. And although they

[rivers] are like this, it is not as if men, when times are quiet, could not provide for them with dikes and dams so that when they rise later, either they go by a canal or their impetus is neither so wanton nor so damaging.

From The Prince

Chapter VII

Studies in Roughness:
A Fractal Primer

THE EARLIEST AND MOST important tools of science come from observing, measuring, and enhancing the output of our senses. The sensation of hot and cold led to the concept of temperature, and from there to the study of thermodynamics. Loudness suggested the decibel scale; pitch and color, the frequency of waves. The experience of heavy and light, fast and slow, underpin the notions of mass and velocity and the study of mechanics. As the great mathematician David Hilbert put it a century ago: "The first and oldest problems in every branch of mathematics spring from experience and are suggested by the world of external phenomena."

But the sensation of roughness had almost entirely been ignored by scientists. Euclid, the Greek geometer whose *Elements* is the world's oldest treatise with near-modern mathematical reasoning, focused on its opposite, smoothness. He and innumerable followers studied smoothness in exquisite detail. Lines, planes, and spheres are the matter of Euclidean geometry, as we are all taught in grade school. I love them; but they are concepts in men's minds and works, not in the irregularity and complexity of nature. How many natural objects around you really fit these old Greek patterns? Maybe the surface of a pond, when there is absolutely no wind or wave, appears truly flat like a plane. Maybe the irises of your children's eyes, if you gaze deeply at them, appear close enough to circular. But how many other smooth, natural things can you name? As I put it in 1982, in my book-length manifesto, *The Fractal Geometry of Nature*: "Clouds are not spheres, mountains are not cones, coastlines are not circles, and bark is not smooth, nor does lightning travel in a straight line."

Now, to talk about fractals and roughness may seem a digression from the workaday task of financial analysis. But a look at the extraordinary range and power of fractal geometry will provide insight into what is possible in finance—and set the stage for further chapters.

The Rules of Roughness

In the past, scientists did their best to view the irregularities of nature as minor imperfections from an idealized shape—like the slight fuzz on an otherwise perfectly smooth peach skin, or the minor distension and dimpling of an otherwise spherical orange. The same assumption stood behind the reasoning of Gauss and Legendre two centuries ago, when they developed the least-squares method of estimating a planetoid's "true," elliptical orbit from a mess of imprecise telescope readings. Once tools like least-squares became available and familiar, other scientists found it easy to follow without much question. For instance, metallurgists used to measure the roughness of a surface or metal fracture by the very same least-squares method—even though they found, puzzlingly, different roughness estimates when measuring different portions of the same metal sample. The same occurs in finance: The "roughness" of a price chart is commonly measured by its volatility—yet that volatility, analysts find, is itself volatile. My contribution was, foremost, to recognize that in turbulence and much else in the real world, roughness is no mere imperfection from some ideal, not just a detail from a gross plan. It is of the very essence of many natural objects—and of economic ones.

More specifically, I developed a geometry that deals with roughness: the mathematical toolkit with which genuine irregularity that goes beyond the fuzziness of a peach can be understood now and, in due time, managed. The key is spotting the regularity inside the irregular, the pattern in the formless. Contrary to popular opinion, mathematics is about simplifying life, not complicating it. A child learns a bag of candies can be shared fairly by counting them out: That is numeracy. She abstracts that notion to dividing a candy bar into equal pieces: arithmetic. Then, she learns to calculate how much cocoa and sugar she will need to make enough chocolate for fifteen friends: algebra. And so it goes in mathematics, from the easiest to the hardest. The fastest way to simplify things is to spot the symmetries, or invariances—the fundamental properties that do not change from one object under study to another.

A fractal has a special kind of invariance or symmetry that relates a whole to its parts: The whole can be broken into smaller parts, each an echo of the whole. Think of a cauliflower: Each floret can be broken off and is, itself, a cauliflower in miniature. Painters, trained to observe nature closely, have known this without waiting for science. Eugène Delacroix remarked, in an article he wrote for *La Revue Britannique*, that

> Swedenborg tells us, in his theory of nature, . . . that the lungs are made of a number of small lungs, the liver of small livers, the spleen of small spleens, etc. . . . Although not being an equally good observer, I still noticed long ago this to be true; I often said that the branches of a tree were themselves complete smaller trees; pieces of rocks are similar to larger rocks, small handfuls of dirt to very much bigger heaps. I am convinced that

many more such examples could be found. A single feather is made of a million feathers.

Fractal geometry is about spotting repeating patterns of this kind, analyzing them, quantifying them and manipulating them; it is a tool of both analysis and synthesis. The pattern can take many forms. It can be a concrete shape that repeats on successively smaller scales, as with the fern or cauliflower. It can be an abstract, statistical pattern—for instance, the probability that a particular square in a grid will be black or white, or that a point in space will be occupied by a star or by vacuum. The pattern can scale up, scale down, and get squeezed, twisted—or both. The way the pattern gets used can be strictly defined by a precise, deterministic rule; or it can be left entirely to chance.

The construction of the simplest fractals starts with a classical geometric object: a triangle, a straight line, a solid ball. That is called the *initiator*. In the last chapter's financial cartoon of the Bachelier model, the initiator was the straight, rising trend line. Then comes the *generator*, or template from which the fractal will be made. That is generally a simple geometric pattern: A zigzag line, a crinkly curve, or—in financial charts—a sequence of prices up $2 last week, down 37 cents today, and up $1.50 the next month. Then comes the process for building the fractal; it is called a *rule of recursion*. For instance, recall how we built the financial Bachelier cartoon. Start with the straight line initiator, squeeze the zigzag generator uniformly in each direction (without turning it) so that its end points coincide with those of the initiator, and then repeat indefinitely. Wherever a straight line appears in the diagram, replace it with a suitably scaled-down copy of the generator. With such fractals, the rules are precise and the outcome predictable, — but also quite elaborate if carried out enough times in enough detail. What had been, in that example, a simple zigzag line, evolves into a jagged curve that, unexpectedly, looks like many natural patterns, for example the profile of a mountain range. In fact, my work convinced computer animators that such fractal processes are the fastest and most realistic tools to draw artificial landscapes and moonscapes.

Fractals get more interesting if you vary the construction process—for instance, reshuffle the straight intervals of a generator in some random order. Or, rather than working with patterns on paper, you could build fractals out of visualized, abstract concepts. Consider social science: The devastating rhythm of war and peace, the unequal distribution of wealth in society, the dominance of big companies in an industry—all can be analyzed as irregular fractal constructs that have more regularity to them than was first assumed. The variety of fractals is immense. But all have a few common traits. First, they *scale* up or down by a specific amount—that is, the parts echo the whole in accordance with a precise, measurable formula.

The simplest fractals scale the same way in all directions, hence are called *self-similar*. They are like high-quality zoom lenses that expand or shrink everything in the frame by the same degree; what they see at one focal length will be similar to what they see at another. But the Bachelier cartoon scales more in one direction than another and the same will be the case with other cartoons of price variation to be introduced in later chapters. Such fractals are called *self-affine*. They are like an office laser photocopier set to shrink a document's image more cross-wise than length-wise. If the fractals scale in many different ways at different points, they are *multifractal*—and their mathematical properties become intricate and powerful. Indeed, the mathematics of fractals in full glory is difficult in its detail. But in its broad strokes, thousands of sixteen- and seventeen-year-olds are now learning it as part of their basic math courses. Fractals are supremely visual, hence supremely intuitive.

Fractals can look haphazard. They often defy conventional geometry or analysis to categorize: They are usually irregular and perplexing, rather than nice and predictable like the parabolas and circles of the old geometers. But the key point: All fractals start simply—simplistically, some might say. In its first stages, any scientific investigation had better be simplistic; otherwise it will never fly. Every fractal is the logical expression of a few straightforward ideas, rules, or mathematical relations. In the simple fractals described here, the initiator, generator, and rule make up the three-letter code for construction, much like the four chemical letters of the genetic alphabet. And as with DNA, so with the fractal code: From this concentrated information come creatures of great beauty and complexity—indeed, of such complexity that sometimes the world's best mathematical minds cannot resolve it.

The distant roots of this field are diverse. It draws upon some oddities first noticed between 1875 and 1925, a fruitful period of turmoil and anarchy in mathematics. They were presented as paradoxes: a line that could completely fill a square, so that, it seemed, one dimension could fill two; an absurdly simple process for converting a solid line into a dust of dimensionless points; an irregular yet continuous curve to which you could not draw a tangent line anywhere. They were fantasies, deliberately contrived to point out some logical inconsistencies in mainstream mathematics. As such, they were initially both advertised and dismissed as monstrous curios, practical jokes on the mathematical graybeards. I increased the variety of these disparate notions manifold. I knitted them into one field, developed it, named it and began applying it to the real world around us—both natural and manmade. It has changed the way many scientists think of the world, even if some poets imagined it first:

So, Nat'ralists observe, a Flea
Hath smaller Fleas that on him prey,
And these have smaller Fleas to bit 'em,
And so proceed ad infinitum.

From Jonathan Swift, On Poetry: A Rhapsody

A Dimension to Measure Roughness

Perhaps the most striking idea in fractal geometry is its peculiar view of dimension. Since Euclid's day, an imaginary mathematical point has had no dimension, a line has had one, a plane, two, and the familiar space we live in, three. Einstein added a fourth, time. Mathematics can generalize the idea, and imagine higher dimensions—purely fictitious, but useful for solving a problem in engineering, economics, or physics. Topology, the mathematical study of surfaces, adds some interesting new twists. From a topological point of view, a cucumber is the same as an orange because one can be remolded into the same shape as the other without having to cut the surface. And the circumference of a circle has the same dimension, one, as a jagged coastline on a shipping map. They are both continuous lines; one can be transformed into the other just by bending, folding and stretching—without cutting.

But is that all there is to dimension? Look at a ball of thread, and think about it first from the idealized viewpoint of Euclid. Assume it is five inches in diameter, made of fiber a fraction of an inch thick. From a long distance away, you can barely see the ball; it is, effectively, a point—of no dimension, according to classical geometry. Hold it in your hand, and it resolves to a normal, three-dimensional ball. Bring it up closer: You see it is a tangle of one-dimensional fibers. Closer still, and the fibers are clearly three-dimensional strands. Keep going until the atoms resolve in an electron microscope: Back to zero-dimensional points again. So what is this ball of thread, anyway? Zero, one, or three dimensions? It depends on your point of view. For a complex natural shape, dimension is relative. It varies with the observer. The same object can have more than one dimension, depending on how you measure it and what you want to do with it. And dimension need not be a whole number; it can be fractional. Now an ancient concept, dimension, becomes thoroughly modern.

Think of dimension, not as an inherent property, but as a tool of measurement. So how do you actually measure something? If you want to measure a straight line, you get a ruler. If you want to measure a curved line, you could use a smaller ruler, inching it along the curve and counting how many times you moved it. You could get a more accurate, if tedious, measure by using a still-smaller ruler; its measurement will be a bit longer than the first, crude one. Eventually, as the ruler keeps shrinking, the measurement settles down to one number that you call the curve's length. But what if the curve is jagged and irregular? What if it is the coast of Scotland? You can start off with a surveyor's glass—a big ruler—and measure from promontory to promontory. Then a long tape might measure point to point. Then a yardstick, then calipers, then a microscope. But this is useless: Unlike the smooth curve, the rocky coastline never provides just one "best" estimate of length. It depends on the scale of the map you want to draw—or your political motives. One researcher, Lewis Fry Richardson, who investigated this paradox nearly a century ago, looked in official references for the surveyed length of political borders between countries. Spanish authorities reckoned their border with Portugal to be 987 kilometers long, whereas the plucky Portuguese counted 1,214 kilometers. The Netherlands measured its border with smaller, poorer Belgium at 380 kilometers, whereas the Belgians counted 449 kilometers.

So how long is it? A useless question, as we have seen. But one way around the problem is to plot on graph paper the measurement you get for each size ruler you use. Of course, the measurements increase as the rulers shrink. But—happy surprise—they often do so at a near-steady rate. Start with a trivial example, a straight line. Say the first ruler you use happens to be exactly the length of the line. Now try a smaller ruler, half as big; it measures the line as two of its lengths. Another ruler, half again as big as the last; the line is four of its lengths. You get the picture. But now try measuring that jagged coastline mentioned earlier. Something unusual develops as you use ever-smaller rulers: The length you measure is growing faster than the rulers are shrinking. And that phenomenon is measured by a quantity called fractal dimension. Begin simply. For a straight line, the fractal dimension is 1. And one dimension is exactly what we expect a straight line to have. But the British coastline, it turns out, has a fractal dimension of about 1.25. Does that make sense? Certainly. A rugged coast is more intricate than a one-dimensional straight line; but however numerous its crags and bays, its outline would not be so intensely convoluted as to fill a two-dimensional square.

That is not all. The Australian coastline, less rugged than the Cornish, turns out to have a fractal dimension of 1.13. By contrast, the smooth South African shore has dimension 1.02, only slightly rougher than a straight line. Another example: rivers. A U.S. Geological Survey study of the course of large American rivers found they have a typical fractal dimension of 1.2 in the East; but in the wilder West, it is 1.4. Again, the measurement fits our intuition of the difference between the rugged Colorado and the placid Charles. Other examples: If you measure the immensely intricate surface area inside the lungs, through which a network of branching bronchia stretch, you find that the total area is vast—something like that of a tennis court. But the fractal dimension is very close to 3. The lining is so convoluted and folded in upon itself that it partakes something of a three-dimensional nature.

What have we here? A new tool to measure, not how long, heavy, hot, or loud something is, but how convoluted and irregular it is. It provides science with its first yardstick for roughness.

Pictorial Essay:
A Fractal Gallery

WITH A SUBJECT as visual as fractals, pictures say more than words. Hence, I offer this pictorial essay on the nature and astonishing variety of fractals, artificial and real.

Literally hundreds of real fractals have been identified. Fractality appears to be part of Nature's basic toolkit—how creatures grow or rocks erode. Why? The answer depends on the context. Consider a rocky coastline again. Physicists speculate that the intricate inlets, promontories, cliffs, and crannies are simply the logical result of wave energy dissipating on a rocky surface. In organic growth, such as lung airways, a process of iterative division is the logical outcome of the genetic rules for animal development: A few instructions, executed simply and repeatedly.

In the 1993 play *Arcadia* by Tom Stoppard, fractal geometry takes center stage. The mathematician-protagonist, Thomasina, tells her young teacher, Septimus:

> Every week I plot your equations dot for dot, x's against y's in all manner of algebraical relation, and every week they draw themselves as commonplace geometry, as if the world of forms were nothing but arcs and angles. God's truth, Septimus, if there is an equation for a curve like a bell, there must be an equation for one like a bluebell, and if a bluebell, why not a rose? Do we believe nature is written in numbers?

Septimus. We do.

Thomasina. Then why do your shapes describe only the shapes of manufacture?

Septimus. I do not know.

Thomasina. Armed thus, God could only make a cabinet.

In fact, fractal structures have also been observed in the work of man, in the pattern of Gothic arches in European cathedrals, in the use of leitmotifs in Wagner's operas, in the skein of paint splashes by Jackson Pollock—even in the frequency and intensity of warfare over five centuries of European history. A superb panorama of fractals can be found on the Yale Web site mentioned earlier, at http://classes.yale.edu/fractals/Panorama/welcome.html. Of course, none of these are conscious products of fractal geometry. But they confirm that it accurately describes some fundamental principles of how people often think and behave: in hierarchies, with repetition and scaling. And after I developed the formal mathematics of it, it began to influence people more directly. Composer Gyorgy Ligeti, among others, has experimented with fractal music. He says:

> Fractals are patterns which occur on many levels. This concept can be applied to any musical parameter. I make melodic fractals, where the pitches of a theme I dream up are used to determine a melodic shape on several levels, in space and time. I make rhythmic fractals, where a set of durations associated with a motive get stretched and compressed and maybe layered on top of each other. I make loudness fractals, where the characteristic loudness of a sound, its envelope shape, is found on several time scales. I even make fractals with the form of a piece, its instrumentation, density, range, and so on. Here I've separated the parameters of music, but in a real piece, all of these things are combined, so you might call it a fractal of fractals.
>
> *From a 1999 interview, The Discovery Channel*

The Sierpinski gasket. Waclaw Sierpinski was a Polish mathematician a century ago who studied, in passing, some peculiar shapes, bizarre constructs that squeeze infinitely long curves inside finite squares. His interest in them was purely theoretical: to challenge some familiar but misleading intuitions of mathematics. He stumbled upon them somewhere, perhaps in decorative designs. After I began my independent fractal researches, I in turn stumbled on this design, brought it to wide notice, and called it a Sierpinski gasket.

It starts with a basic shape called the initiator—in this case, a black triangle at top left. Think of it as the canvas on which the fractal drawing will start. Immediately beside it comes the generator, or template for building the fractal. In this case, the generator is the original triangle that was first shrunk to half in both height and width, and then cloned three times to fit inside the original black triangle. At bottom left come the instructions for completing the drawing. Replace each solid triangle with an appropriately scaled-down version of the generator. If you keep repeating the process, over and over again at ever-smaller scale, you get the pattern shown at right: lacy and insubstantial.

The fractal skewed web. Fractals can fit into any dimension—even our familiar three. This one, with perspective added, begins much the same way as did the Sierpinski gasket. Instead of a triangle, we now have a set of stacked tetrahedrons, or pyramids. Eiffel designed his famous tower in Paris using trusses arranged in what we would now call a fractal pattern. The design yields the greatest strength for the least steel.

This and the preceding diagram exhibit *self-similarity,* a property common to many of the simplest fractals. At every scale you look, each element of the diagram is similar in shape to the element on the next scale higher up or lower down; "similar" means reduced in size with no deformation. Finance requires a different class of fractals called *self-affine,* meaning that the scaling happens faster horizontally than vertically. In more general fractals, the parts can get systematically twisted, rotated, or in other ways transformed.

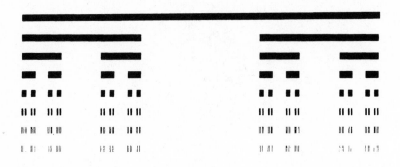

The Cantor dust. This is one of the oldest fractals, named after Georg Cantor, a Russian-German of the nineteenth century who radically changed the way mathematicians think about infinity, sets, and many other basic ideas previously taken for granted. The Cantor dust is typical of his paradoxes. It starts as a simple line: straight, continuous, and one-dimensional (here a thickened bar to make it possible to actually see it). Its generator is the same line with the middle third punched out. The rule: Keep replacing the ever-shorter lines with ever-more porous generators. The result, if kept up forever, is totally unexpected. No solid bit of line is left anywhere. All that remains is an irregularly spaced sprinkling of individual points. I call this process *fractal curdling,* after the way the clumps of heavy curd in whole milk settle out from the whey.

Cantor believed he was fleeing away from Nature, but Nature appears to be fond of his construction. The rings of Saturn are a collection of concentric near-circles, diaphanous and passing sunlight. They are spaced irregularly, as if a Cantor dust had been dragged like a broken-toothed comb around a vast circle centered at the planet's core, and the dust of space had settled in the resulting grooves. On earth researchers have found that the spectra, or energy "fingerprints," of some organic chemicals resemble a Cantor dust.

The Koch curve. In 1905, Swedish mathematician Helge von Koch described a construction that recalled a snowflake, with jagged edges and symmetrical forms. Like Cantor's dust and Sierpinski's gasket, its intent was to defy conventional mathematical notions. Its outline is truly monstrous: continuous but of infinite length; you could not draw a line that was tangent to it anywhere along its infinite length. This sort of mathematical anarchy annoyed many contemporaries, who were still pursuing ideals of continuity and order. A French mathematician, Charles Hermite, wrote in 1893 of "turning away in fear and horror from this lamentable plague of functions with no derivatives."

The Koch curve is one-third of the snowflake. As with the Cantor dust, its construction starts with a straight line—here shown as the horizontal side of the top triangle. But instead of deleting the middle third, you push it out to form a triangular tent over the mid-section. As shown down the left side, the fractal is formed by replacing each of the ever-shorter intervals of a broken line with ever-smaller versions of the tent generator. Soon a paradox emerges. Each repetition adds more tents, turning a short, straight line into a jagged trail that is longer than the original in the ratio of four-thirds. The length of the curve grows and grows.

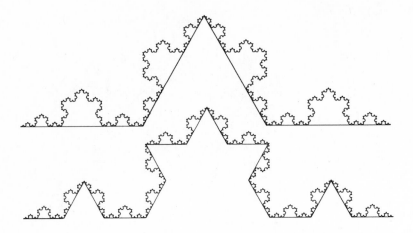

Fractal dimension. One of the most significant concepts in fractal geometry is dimension, a numerical measure of the "roughness" of an object. We are familiar with the one dimension of a straight line, or the two dimensions of a plane—but how about a fractional dimension between the two?

Look at the Koch curve above, and try to measure its length. Start with a ruler one-third the object's breadth. That is the triangular line fitting inside the curve, in the top panel. As you can see, it fits four times. Then shrink the ruler by a third, as in the bottom diagram. Because it can now fit into more crannies of the curve, it measures more distance—in fact, four-thirds as much. Continue the process, shrinking the ruler and measuring. At each stage the length measured is multiplied by the same ratio: 4 to 3. The fractal dimension is defined as the ratio of the logarithm of 4 to the logarithm of 3. A pocket calculator converts that: 1.2618. . . . This makes intuitive sense. The curve is crinkly, so it fills more space than would a one-dimensional straight line; yet it does not completely fill the two-dimensional plane.

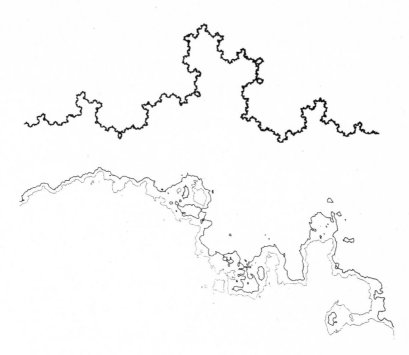

Random fractal curves. So far, all the fractals in this gallery have been regular and, once you knew the rule, the constructions were exactly repeatable and the results, predictable. But such constructions are nothing but appetizers. I like to call them cartoons. Adding an element of chance complicates the game, and starts to produce structures that look more like sports of Nature than of man.

The top diagram is the Koch curve again, with luck added. It starts with the same initiator and generator as shown earlier. But whereas the prototypical Koch curve plugs the ever-shrinking generators in exactly the same way at each step, here we toss a coin at each step to decide whether to place the "tent" right side up, or upside down. The result is more irregular and flows more naturally. In fact, it starts to look a bit like a coastline. The bottom diagram, using a more complicated fractal process driven by a computer, starts to look startlingly real—as if traced from a shipping chart.

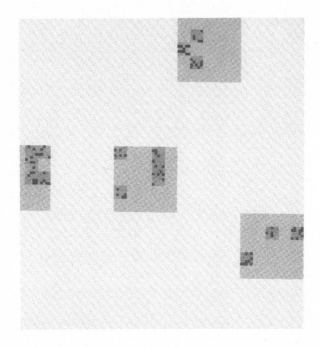

Random fractal dusts. These diagrams illustrate another face of
randomness. It remains an unsolved problem of cosmology to fully
describe and explain the irregular distribution of stars and galaxies
across space. Of course, it is known that the stars coalesced, by force of
gravity, as did the galaxies. But exactly how they ended up in their current positions is unknown. I have proposed a fractal scenario. The diagram above presents a random fractal obtained by random curdling.
Start with a big square, divide it into 125 smaller squares, and choose a

random number of them to darken. Repeat the process in each of the darkened squares. The result, after many iterations, is a faint dusting of black points, in the diagram on this page. To an astronomer, it resembles a diagram of galaxy clusters. On the completed diagram (above), you cannot immediately spot any fractal recursive process at work. But it is present, and computer analysis would reveal it to be so. Such is the power of fractals and chance working together: Simple rules build complex structures, and complex structures deconstruct into simple rules.

Fractals in the physical world: clouds and cluster. With random processes added, we finally start to see the hand of nature. The top diagram is the work of a computer to illustrate the principle. It represents a completely artificial cloudy sky. The bottom diagram illustrates fractal growth starting from an irregular "seed" in the center. As a random fractal process adds particles to it step by step, tendrils and branching structures slowly appear to yield a structure called DLA: a diffusion-limited aggregate, one of the most fascinating, ubiquitous, and difficult objects of statistical physics.

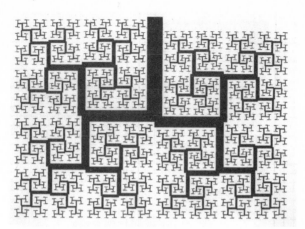

Fractals close to home. On top is a computer-drawn fractal cartoon; it uses an irregular, branching structure as generator. Below (Weibel 1963), the natural prototype for this: the complex branching of bronchia inside a human lung. In fetal development, the lungs form step by step. The bronchial tubes branch. Those branches in turn branch again. And so on, down twenty-odd levels of branching from large tubes to very small, according to anatomical studies. The outcome: a fractal space-filling sponge of lung tissue, with convoluted, branching airways providing oxygen in precisely regulated volume and velocity to millions of tiny air sacs.

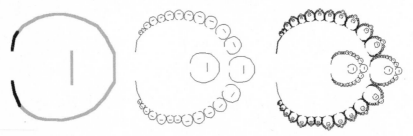

Fractals in society. Economics, anthropology, history, musicology, architecture—the list of social sciences and humanities in which fractals have been found is long, indeed. The photograph (American Geographic Institute) is an aerial view of the Ba-ili settlements of southern Zambia. It is an enclosure to pen in livestock; but it is formed of dwellings, side-by-side, in a ring. The larger the dwelling, the more important the family until, in the center, is the chief's house. Finally, within each dwelling is a household altar. The diagram, from Eglash 1999, shows how the village follows a fractal hierarchy.

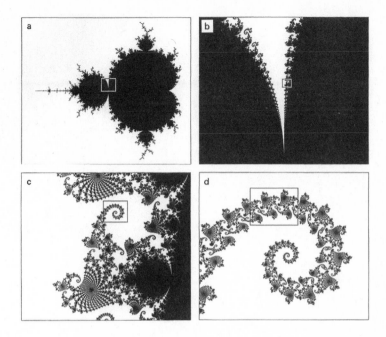

Chaos and the Mandelbrot set. To conclude this pictorial essay, the last example is perhaps the most famous one: a mathematical chimera that my colleagues named for me. Since I discovered it over two decades ago, it has become the topic of innumerable mathematical investigations. It has also been reproduced on millions of T-shirts, prints, book jackets, and PC screen savers. Readers seeking more precise explanations of it may consult my recent book, Mandelbrot 2004a.

The Mandelbrot Set illustrates the profound connections between fractal geometry and chaos theory. It uses a remarkably simple mathematical feedback loop to produce an astonishing variety and complexity of results. As you zoom in on any portion of it, as if with a microscope, the pattern does not get simpler as you would normally expect. "Proper fractals" remain equally complicated at every level of magnification. But the Mandelbrot set's complication increases without any bound. Its perplexing mix of simplicity and complexity has made it a mathematical Everest, and two mathematicians who did not reach the summit but went far up some of its faces won Fields Medals sometimes called the mathematical equivalent of the Nobel Prize, for their labors.

CHAPTER VIII

The Mystery
of Cotton

MOMENTS OF DISCOVERY—the "eureka" moments when a scientist leaps out of the bathtub with a new law of physics clear in his mind—are rare. More often, discovery is a long, tortuous path signposted by more questions than answers. Certainly, my main discoveries have often begun with a mystery, and, at that, a mystery of a special kind: a mental gridlock of sorts between some established theory of science and new data that challenge it. Thus it was with one of the central mysteries of finance.

It was 1961. I had been working a few years at IBM's main laboratory up the Hudson River from Manhattan. It was a surprising place for a pure scientist. The company had re-tooled itself from a manufacturer of mechanical tabulating machines to a pioneer of electronic computers; and for that task, it had staffed up a large laboratory by including a number of brilliant misfits who were allowed to pursue every imaginable topic. Some were obviously related to computers, but many not. I, a recent arrival from France, was working on a new use for computers: economics. I was studying reams of computerized data, analyzing how income gets distributed through a society—the proportion of rich to poor, of superrich to very rich. My work intrigued a few economists in the world outside, and so it was that I was invited to Harvard one day to give a talk.

I arrived there to find a surprise. In the office of my host, Professor Hendrik S. Houthakker, I spotted a diagram on his blackboard. It had a peculiar, convex shape—a kind of "V" turned to open to the right instead of the top. It was nearly the same shape as a diagram on income distribution that I was about to draw for my lecture. How was it, I asked, that something like my diagram was on his wall?

He looked at me blankly. "What do you mean? I have no idea what you're going to talk about." His diagram was not on income, but on cotton prices. He had been working with a student before I arrived; and the blackboard had not yet been erased.

Now here was a puzzle that grabbed me. Why would a diagram on the way rich and poor spread through society look like one on how cotton prices buck up and down? Was it pure, dumb coincidence? Their bizarre convexity struck me. Could it reveal some deeper connection between the two—some odd truth lurking behind the charts? Houthakker, it emerged, had been studying cotton prices for a while, getting nowhere. Mainstream economists had, by then, rediscovered Bachelier's hypothesis about how prices vary as if by the toss of a coin. They were beginning to plow through price records, looking for evidence. At the time, reliable, long-running records of commodity or security prices were hard to come by. Cotton was an exception. For more than a century, the New York Cotton Exchange had kept exacting, daily records of prices as the vital commodity moved from the plantations of the Old South to the dark mills of the industrial North. Virtually all interstate trading was centralized at one exchange. It was a huge, liquid market, with ample resources for record-keeping. Such old, accurate, centralized price data should have been an economist's dream. But for Houthakker they proved a nightmare. No matter how he manhandled the numbers, he could not get them to fit Bachelier's model. There were too many big price jumps and falls. And the standard measure of how much they varied—the volatility, or standard deviation—kept shifting over time. Some years prices were stable, other years wild. None of his statistical tools could resolve the muddle.

"I've had enough," he told me. "I've done everything I could to make sense of these cotton prices. I try to measure the volatility. It changes all the time. Everything changes. Nothing is constant. It's a mess of the worst kind." In short order, we made a deal. I would take over the cotton prices. He handed me cardboard boxes of computer punch-cards, to which the prices had been transferred. If I could make sense of them, good luck.

Back in New York with Houthakker's boxes, I asked the IBM computing center to assign a programmer to me. I would take the visual cue from that Harvard blackboard. A computer program would analyze the cotton price records, just as it had done for the income records: How many big price jumps, how many small; how many big incomes, how many small? While I waited for the results—a long wait, given my own low standing on the computing center's priorities list—I took the commuter train into Manhattan, where the National Bureau of Economic Research was then located. In its library I found many books, covered in dust and filled with tables of financial data—a treasure in 1961, though a meager haul by today's data-swamped standards. Later, I needed more cotton price records and wrote to the U.S. Agriculture Department in Washington. I gathered every available datum. I built an encyclopedia of cotton prices daily, weekly, monthly, and annually over more than a century. And the computers helped look for the patterns.

What they found was extraordinary—in fact, the 1963 paper describing my work, "The variation of certain speculative prices," became one of the most frequently cited in the economics literature. It sparked great controversy. It yielded my theory of the first of two fundamental aspects of how financial prices behave; as will be seen, subsequent research added new perspectives. But the winding trail by which I came to these discoveries became, in itself, a part of the story of fractal finance. Solving the cotton mystery required at least three broad strands of thinking to come together—on power laws, on the distribution of personal income, and on a then-esoteric topic in what seemed a totally different universe, the mathematics of stable distributions. As will be seen in subsequent chapters, a second conundrum, on the floods of the Nile River, led to a discovery concerning a different fundamental aspect of finance. And, after an initial attempt in 1972, in the late 1990s I finally pulled together all these different clues into one, comprehensive solution to the mystery of financial theory.

But like all good detective stories, this one begins with the smallest of clues, ignored by most. In fact, it had been discarded—quite literally.

Clue No. 1: A Power Law Out of the Blue

In 1950, I was a young mathematics student in search of a good topic for my thesis at the University of Paris. My uncle, Szolem, was the local model of a mathematics professor: Very theoretical, deeply conservative, and—despite his having been born in Poland—a pillar of the French academic establishment. He was elected to a full professorship at the prestigious College de France at the precocious age of thirty-nine.

This was the era of Bourbaki, a mathematical "club" that, like Dada in art or Existentialism in literature, spread from France to become, for a time, hugely influential on the world stage. It worshipped abstraction and math for math's sake; it scorned pragmatism, applications, or math as a tool for science. It was the dogma of French mathematics, and a reason why I ultimately left Paris for IBM. I was a young rebel, much to my uncle's consternation. While

pany. His first wife was a Russian countess; she left him for a young servant. Pareto did not begin serious work in economics until his mid-forties, but he swiftly made a mark and settled in Lausanne, Switzerland, as a professor and scholar. He started his career a fiery liberal, besting the most ardent British liberals with his attacks on any form of government intervention in the free market. He ended as, if not a believer, at least a student of socialism. He died in 1923 among a menagerie of cats that he and his French lover kept in their villa near Geneva; the local divorce laws—he was still officially yoked to his fickle countess—prevented him from re-marrying until just a few months before his death. His legacy as an economist was profound. Partly because of him, the field evolved from a branch of social philosophy as practiced by Adam Smith into a data-intensive field of scientific research and mathematical equations. His books look more like modern economics than most any other texts of that day: tables of statistics from across the world and ages, rows of integral signs and equations, intricate charts and graphs.

One of Pareto's equations achieved special prominence, and controversy. He was fascinated by the problems of power and wealth. How do people get it? How is it distributed around society? How do those who have it use it? The gulf between rich and poor has always been part of the human condition, but Pareto resolved to measure it. He gathered reams of data on wealth and income through different centuries, through different countries: the tax records of Basel, Switzerland, from 1454 and from Augsburg, Germany, in 1471, 1498, and 1512; contemporary rental income from Paris; personal income from Britain, Prussia, Saxony, Ireland, Italy, Peru. What he found—or thought he found—was striking. When he plotted the data on graph paper, with income level on one axis and number of people with that income on the other, he saw the same picture nearly everywhere in every era. Society was not a "social pyramid" with the proportion of rich to poor sloping gently from one class to the next. Instead, it was more of a "social arrow"— very fat at the bottom where the mass of men live, and very thin at

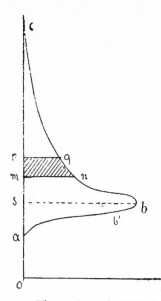

The income curve. This is Pareto's 1909 diagram of how wealth is distributed through any human society, in any age or country. Rising income is on the vertical scale, population on the horizontal (latter-day economists have switched the coordinates). The number of people with income between levels *m* and *p* is represented by the shaded area. The mass of men fall to the broad bottom of the curve. The privileged few sit at the narrow top. While the bell curve is symmetric, the income curve is not.

the top where sit the wealthy elite. Nor was this effect by chance; the data did not remotely fit a bell curve, as one would expect if wealth were distributed randomly. It is a social law, he wrote: something "in the nature of man."

That something, though expressed in a neat equation, is harsh and Darwinian, in Pareto's view. At the very bottom of the wealth curve, he wrote, men and women starve and children die young. In the broad middle of the curve all is turmoil and motion: people rising and falling, climbing by talent or luck and falling by alcoholism, tuberculosis, or other forms of unfitness. At the very narrow top sit the elite of the elite, who control wealth and power for a time— until they are unseated through revolution or upheaval by a new aristocratic class. There is no progress in human history. Democracy is a fraud. Human nature is primitive, emotional, unyielding. The smarter, abler, stronger, and shrewder take the lion's share. The weak starve, lest society become degenerate: One can, Pareto wrote, "compare the social body to the human body, which will promptly perish if prevented from eliminating toxins." Inflammatory stuff—and it burned Pareto's reputation. At his death in 1923, Italian fascists were beatifying him, republicans demonizing him. British philosopher Karl Popper called him the "theoretician of totalitarianism."

By the time I heard of him, much of the fire had gone out of the debate. Most economists willingly adopted his seminal theories on other topics such as economic equilibrium. But they passed over in silence the distasteful matter of his income curve. To me—I did not even study economics until I was in my thirties—Pareto's formula was a marvel.

He grouped people by their incomes, counted how many were in each category, and then plotted the results. Now, it is a handy fact about data that scale according to a power law: On suitably chosen paper, they form an unmistakable pattern. Start with some engineering graph paper, of the kind that has logarithmic measures on each side; that is, instead of numbering the axis scales 1, 2, 3, or 4, call them 1, 10, 100, and 1,000 units, in powers of ten. On paper like this, scaling data will form a straight, sloping line; other data will not. You can try it yourself. Get a sheet (available on the Web, if you do not keep it handy), and plot the area of an assortment of square bathroom tiles. Call the horizontal axis the length of each tile, and the vertical axis the area. Then plot: a tile of length two inches has an area of four square inches, a tile of length three inches has an area of nine square inches, and so on. A straight line will emerge, climbing from left to right. How fast does it climb? Measure the slope. It rises two units for every one unit sideways. Slope: 2. Funny coincidence: Two is also the value of the exponent by which you raise the length to get the area. In short, the slope of the line is also the "power" in the power law. It works with other powers, too. If you fill a boxcar with cubic boxes, the volume increases by the power of three and the slope will be steeper. If you create a long string by lining up shorter strings end to end, the power is one.

Of course, bathroom tiles, boxcars, and strings make for particularly silly power laws; other, more complex data may show steeper or shallower slopes on the paper. Regardless: If a power law is in play, some kind of straight line will appear. It is a simple test, childishly simple. Just draw, see, and measure. The diagram below shows some examples.

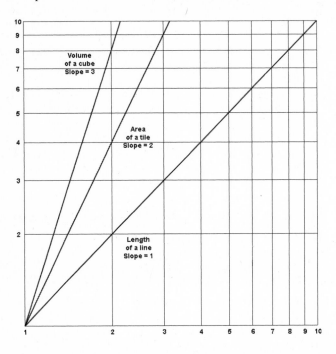

The picture of an increasing power law. This is what a power law, such as Pareto investigated, looks like when plotted on log-log paper (with the axes measured off in powers of 10). The slope of the line equals the "power" in the equation. For instance, the area of bathroom tiles rises as the square (power = 2) of the length of each side of the tiles. The volume of boxes rises as the cube (power = 3) of the side. And the length of a line rises in direct proportion (power = 1) with the length of the shorter lines laid end to end to form it. With fractal geometry, one can get an endless range of powers—that is, diagrams that slope more steeply or gently, or go down rather than up.

A straight line is exactly what Pareto found, when he plotted income against the number of people. A power law was clearly present. In fact, his line sloped down instead of up, because the power was negative rather than positive. And alpha, Pareto's name for the absolute slope of that line, was 3/2, he thought. What does that mean? Well, the gentler the slope, the more even the distribution of income. In the case of human vocabulary, Zipf thought it was governed by a power-law slope of 1: With so gentle a slope, the average person would use a few words very often but still have a fairly rich vocabulary overall. (Zipf happened to be paying undue attention to James Joyce's *Ulysses*. Most books show a slope greater than 1—that is, they have a less rich vocabulary.) With a Pareto slope, or alpha of 3/2, much wealth is concentrated in very few hands.

Look at some specifics. Pick a group of people to study—say, everybody making more than the U.S. government's $5.15 minimum hourly wage, or $10,712 a year. Now ask: What percentage of people earn at least ten times that? According to Pareto's formula, the answer should be 3.2 percent. Now go higher up the moneyed classes: What proportion of those above minimum wage is earning more than $1.07 million? Answer: 0.1 percent. And once more: What proportion earns more than $10.7 million, a thousand times the minimum? Answer: 0.003 percent—a very small number, indeed. Look at it another way, through the lens of what mathematicians call conditional probability. That is a fancy term for a straightforward concept: Given a starting condition, what is the probability that some event will happen? The absolute odds of being a billionaire are very low; but according to Pareto's formula, the conditional probability of making a billion dollars once you have made half a billion is the same as that of making a million once you have made half a million. Money begets money, power makes power. Unfair, but true—both socially and mathematically.

As it turned out, Pareto's calculations were hobbled by the limitations of his data. His formula works only when looking at the very rich. He was also handicapped by his excessive hope of finding a universal law, for all countries and ages. Just as Zipf thought that for word frequencies alpha is always 1—which it is not—Pareto thought that for income it was the same in every state—which it is not, either. In most cases, he underestimated alpha, which appears to be closer to 2 than 3/2—meaning millionaires are rarer than he thought. But his basic observation of a power-law relationship between income and population was insightful. At its core is the observation that, in a society, a very few people are outrageously rich, a small number are very rich, and the vast bulk of people are middling or poor. The alpha in Pareto's formula is just a way of quantifying exactly how inequitable the society is.

And it permits some interesting predictions. For instance, some economists have found Pareto's formula a good way to describe incomes within individual professions—say, the pay scale in the electrical industry from executive suite to turbine room. From that, I developed a mathematical model of why people often specialize in one profession, rather than hop from trade to trade. The reason is common sense, but the mathematics support it. If they invest in their own profession—for instance, get a graduate degree—they will probably rise higher and faster up their own industry's income curve. If they change fields or dabble in many, they will probably make less. That helps explain why, when a new multidisciplinary industry like e-business appears, salaries can shoot up: The new companies have to offer absurdly high compensation to induce people to take the risk and leave their own, specialized Pareto curve.

In different guises, power laws like Pareto's occur throughout economics. For example, data suggest that the size of firms in an industry scales. The bigger the company, the rarer it is, in a proportion that follows a Pareto-like formula. City sizes in a country scale, as well, from metropolis to village. Insurance claims make a particularly good and well-accepted example. In Sweden the damage claims for house fires are collected by an actuarial agency of the government. It was found that the odds of any particular size of claim vary much like income—except that in insurance, alpha for most

houses is about ½. In a sense, that means fat insurance claims are relatively more common than fat millionaires. In Pareto's case, just
0.003 percent of people had incomes more than 1,000 times the minimum. But with Swedish house fires, damages 1,000 times the
deductible would account for 3.2 percent of all claims. For an insurance company, this is not a trivial difference; it helps show the
importance of the exponent, alpha. By necessity, insurance companies are very familiar with power laws. Denying them would create
an additional and totally unnecessary risk. Same formula, different
result because of one change in the parameters. It is all exquisitely
versatile.

Clue No. 3: The Laws of Exceptional Chance

The last hint in the cotton mystery goes back again to my student
days. After the war, I was at the École Polytechnique, one of France's
"Ivy League," the Grandes Écoles. One of my professors was Paul
Lévy, a well-known mathematician, and the same man who had
unintentionally played so decisive a role in Bachelier's life story.

Lévy was independently wealthy, the scion of a Jewish merchant
and academic family. To students at the back of his lecture hall—as
I was—he was near-inaudible and his long, gray, and well-groomed
figure bore an odd resemblance to the somewhat peculiar way he
had of tracing the long "?" of an integration symbol on the blackboard. He was a misfit, as I, myself, was fated to become: a member
of no club, movement, or establishment. Though now acknowledged as one of the greatest probabilists, he was at the time largely
ignored by other French mathematicians. That was partly his own
fault: He was notoriously sloppy in his written proofs and scientific
publications, making careless errors from haste that haunted him
after. Some of his most unusual ideas he never published; they
seemed, he later said, too obvious—though others who stumbled on

the same notions and did publish achieved recognition. He was suf-
fered to give occasional lecture series at the University of Paris; it
was feared he would in some way disrupt the standard curriculum.
I recall that by the end of one such series, I was his sole auditor; we
could as easily have quit the auditorium and adjourned to his office
for a chat. At seventy-eight, he received belated recognition by elec-
tion to France's Académie des Sciences. But he was ever an anomaly.
As a later teacher of mine, John von Neumann, told me: "I think I
understand how every other mathematician operates, but Lévy is
like a visitor from a strange planet. He seems to have his own private
methods of arriving at the truth, which leave me ill at ease."

Lévy did not "arrive at" probability theory until he was nearly
forty, when he was asked shortly after World War I to lecture on
targeting errors in gunnery. He was soon doing original work,
beginning with what he—most unfortunately—called "stable"
probability distributions. Now, stable means that you can do something to an object—for instance, rotate it, shrink it, or add it to
something else—and its basic properties remain unaltered. A
Gaussian bell curve is stable in this sense. For instance, the theory of
errors assumes that every kind of error of measurement traces a bell
curve. And it is stable: You can add the errors of measurement coming from two independent sources, and the combined data set will
still trace a bell curve. The average may have changed, or the standard deviation may have widened; but it remains a bell, nonetheless.
Oddly enough, as Cauchy observed long ago, the same thing happens with his distribution, the blindfolded archer described earlier.
If you add the target scores of the blindfolded archer with those of,
say, a blinded gunner, the two sets of data together will still fit
Cauchy's formula. It, too, is stable. In fact, there is a whole family of
such probability distributions. I called them "L-stable," in honor—
and later in memory—of Lévy.

What distinguishes one family member from another is the relative importance of the largest individual measurements. Recall: If you
add the heights of 1,000 people and calculate the average, adding a

1,001st person's height will not change the average very much. With the blindfolded archer, by contrast, the 1,001st shot, if very wide of the mark, could totally change the average. The bell curve is egalitarian; every data point adds its value to the whole, but no one can dictate the statistical outcome to the rest. The Cauchy curve is inequitable and dictatorial; the big data points can and do dominate the crowd. These are two extremes, and Lévy linked them by a whole spectrum of other family members. All can be expressed by the same basic formula. Only the details—the parameters, in mathspeak—differ. If you fiddle with the parameters, you get curves that are squatter or taller, have more outliers or fewer, are shifted left or right as the median changes, and are symmetrical or skewed. The key parameter is alpha, the same variable as in Pareto's and Zipf's formulas.

So, many seemingly unrelated ideas come together in one unifying concept. This sort of serendipity, to a mathematician, is better than winning the lottery on your birthday. Of course, Lévy's interest in this topic was strictly theoretical; and his early discussions of stable distributions called them "exceptional." He was merely picking up a thread of math that had begun with the practically minded Gauss, Poisson, and Cauchy and had been revived in "purified" form with George Pólya, Kolmogorov, and Lévy, himself. Lévy eschewed applications of any kind. When I began studying income it occurred to me that these math games might actually be useful: Alpha and all the other details of stable distribution theory might make handy tools for analyzing the real world. That proved to be the case, as I showed in several papers on the distribution of personal income. But it was initially difficult to get my work accepted—by either applied or theoretically minded scientists. The applied researchers found the math of L-stable distributions formidable, especially the way they do not behave "properly" with an easily defined variance. They recall the difficulty calculating the score of our blindfolded archer. Theoretical scientists were simply not interested. When I later told Lévy how I had developed his ideas and applied them to economics, he was flabbergasted and, perhaps,

annoyed. In his view, "real" mathematicians simply did not do such
prosaic things as study income or cotton prices.

The Cotton Case: Basically Closed

So, three clues: Power laws, the spread of rich and poor, and the
mathematics of the exceptional stable probability distributions. One
is a general way of looking at the world; the next a practical example of it in economics; and the last a set of mathematical considera-
tions no one viewed as useful. How did they come together in the
cotton mystery?

The IBM computing center chewed through the thousands of
cotton prices, as I had asked. It quickly confirmed Houthakker's
view: The price changes from one day to the next, one week or
month or year to the next, did not behave as the Bachelier model
assumed. The variance misbehaved. Each time I added an extra
price-change to the data set, my estimate of the variance changed. It
never settled down to one simple number—say, 1 percent volatility.
Instead, it roamed erratically from about 0.4 percent to 3 percent,
nearly a tenfold difference. That was surprising, considering that
the quality of the data itself could not be challenged. Moreover,
there were too many big price jumps to fit the bell curve.

This was indeed a problem. Clearly, I surmised, a power law was
at work—just as in Zipf's word frequencies and Pareto's income
curve. The size of price changes varied in the same way. A great
many small price movements are found in the same cotton market
with a few enormous jumps; a great many rare words are in the dic-
tionary with a small number of common words; vast legions of poor
people coexist in the world with a privileged few plutocrats.
Uneven. Unfair, perhaps. But still indisputable.

That was part of my hypothesis. How to test it? Well, if I was
right, then I should be able to find a particular value of alpha that
governs the cotton price curve—just as Pareto thought he had

found an alpha of 3/2 for income. So I followed Pareto's lead and drew a diagram for cotton prices on log-log paper. Seeing the plots, at last, was satisfying. Every kind of cotton data that I considered formed a straight line. Together they form the "rows of cotton" diagram that follows. The fits to the line are not precise; in statistics, nothing is. But if you hold a ruler against the lines, you can measure the slope: It is -1.7. It is negative because the line falls, rather than climbs; by convention, the alpha would be called 1.7. Cauchy's or Zipf's distributions would have a smaller alpha, of 1, Pareto's, 1.5, and the profit of a coin-toss game, 2. So the variation of cotton prices

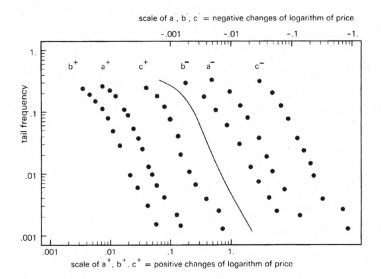

Rows of cotton. The easiest test of whether data scale is to plot them on some engineering graph paper—the kind like this with logarithmic scales in both directions. If the data form a straight line, then some kind of scaling is at work. That is what this somewhat complicated chart, from my 1963 paper analyzing a century of U.S. cotton prices, shows. For each of three sets of cotton data, I analyzed the positive and negative changes separately. The results are shown in these six rows of dotted lines. The horizontal scales show the size of the price changes; big ones are to the right. The vertical scale shows the frequency of each change; the most common are to the top. Most of the dots slope about the same way, as shown by the solid line. This is *prima facie* evidence of a power-law behavior and (digging deeper) an L-stable distribution.

fits somewhere between that of a working man's income and a gambler's winnings. Surely, there is something poetic in that fact.

But the cotton diagram has a deeper story to tell. Time is the crucial dimension to consider. If you say the price per bale fell one dollar, what time-scale do you mean? With the income curve, time was irrelevant; Pareto had taken a snapshot of yearly incomes of a collection of individuals not ordered in any way. The cotton market is a movie, varying over time. Normally, we think about it from one celluloid frame to the next, from one trading day to another. But we could as easily edit it to view it only once every twenty frames (a month of trading days) or once every 250 frames (a year of trading days). Now, you would typically expect the movie to look different for each edit. But what if it looked the same? That was my idea: What the heck? I decided to check whether by any chance the movie looks roughly the same in all three cases. If it did, at each time-scale you should see the same proportion of big changes to small, the same fat tails, the same odds of another big change coming. To test this invariance, look back at the cotton chart. The different data sets shown are actually different ways of playing the movie. The ones labeled b^+ and b^- show the way prices varied daily from 1944 to 1958 (b^+ shows price rises, b^- the price falls). The ones labeled a^+ and $a-$ show the daily variations from an earlier period, 1900 to 1945. The third set, c^+ and c^-, shows monthly price changes from 1888 to 1940.

All look the same. A month looks like a day, one set of days like another. In fact, at a first approximation, you could not readily tell without the labels which line was which. That clicks with something else. Having acquired an interest in financial markets after my move to New York, I started chatting with the Wall Street pros. There is something funny, one told me: In the newspaper, all price charts look alike. Sure, some go up; some down. But daily, monthly, annually—there is no big difference in the overall look of it. Strip off the dates and price markers, and you could not tell which was which. They were all equally wiggly.

"Wiggly" is hardly a scientific term—and until I developed fractal geometry years later, there was no good way to quantify so vague a notion as wiggly. But that is exactly what we can now see in the cotton data: a fractal pattern. Here, the fractal scaling up and down is not being done to a shape—the florets of a broccoli or the triangles of a Sierpinski gasket. Rather, it is being applied to a different sort of pattern, the way prices vary. The very heart of finance is fractal.

So it all comes full circle. It was no coincidence that Houthakker's cotton chart looked like my income chart. The math was the same.

The Dénouement

Like most trade unions, economics departments like to keep a closed shop. So my cotton research caused a hullabaloo.

My initial paper on the subject started life early in 1962 as a messy draft that I dashed off, pecking with two fingers on a portable typewriter. But Houthakker, it happened, needed a last-minute replacement for a Harvard colleague who was going on sabbatical; he asked me to teach in his place. My draft, warts and all, was hastily published as an internal research report by IBM. The noise from academia was loud. Who was this Mandelbrot fellow, a grimy industrial scientist with a degree in applied mathematics, to challenge the elaborate models of the economics elite? But curiosity was strong and spreading. On a stopover at Chicago, I met economist Zvi Griliches (who later moved to Harvard). He was setting the program for a winter meeting in Pittsburgh of the Econometric Society. I was invited to speak, and a panel of three was assigned to lead a discussion on the cotton results. Then, while elsewhere at Harvard late in 1963, I got a call from across town at MIT's Sloan School. Cootner, an economist there and a discussant in Pittsburgh, was preparing a book to be called *The Random Character of Stock Market Prices,* a compendium of academic views on the mathematics of markets, beginning with a translation of Bachelier's thesis. He said he wanted to feature my work, but all the other articles in the book were reprints from proper academic sources, not something from an in-house corporate press. Could I get it into print somewhere, anywhere, in time for his deadline?

I called every economics journal around. Some asked me to spell my name; others asked about my background. A few knew me, but said the deadline was too short; economics journals are notorious in the academic world for sitting on publishable research for months or years, before leisurely trickling it out. My luck turned with the *Journal of Business* at the University of Chicago—which, ironically, was to become home to the most ardent proponents of the standard financial model. An editor there, Merton Miller, later a Nobel laureate, took a few hours to check around, and then called back with a deal. My IBM paper, he said, was already well-known around his economics department. In fact, my ex-student Fama was on the faculty. Therefore they could skip the usual, time-consuming process of having a paper read, critiqued, and dissected by academic "referees." I jumped at the offer. I mailed a rough copy to Miller within the week; publication of the current issue was postponed; and space was made for my paper by postponing a less "hot" property. Page proofs came in no time. Cootner's condition of prior academic publication was duly met. Fama edited my text and wrote an introduction for it. This was a helpful translation, for economists, of what I was trying to tell them as a mathematician. I have since found translators helpful when selling new ideas in a hostile marketplace.

And the reception was hostile. Following a reprint of Fama's introduction to my paper, Cootner included a five-page critique in his book. He felt my graph-paper test too simplistic, the math intractable, the evidence insufficient, and cotton too peculiar a commodity from which to draw sweeping conclusions. The implications were great, he wrote. But "surely, before consigning centuries of work to the ash pile, we should like to have some assurance that all our work is truly useless."

Assurance was not long in coming. My own research found scaling patterns in the shares of such nineteenth-century railroad Goliaths as the B&O, the Boston & Maine, and the Illinois Central. Fama and his students at Chicago found more evidence. In 1970, for instance, Richard Roll found U.S. Treasury bill yields scaled with a pattern more erratic than that of cotton prices. There were more. As Fama put it at one point, there "would seem to be conclusive evidence in favor of the Mandelbrot hypothesis."

Economics is a science of fashions—Keynes and "pump-priming" at one time, Friedman and monetarism at another. The profession burns through new theories the way a teenager hops from one new date to another: It meets them, spends some time with them, examines them, finds what it thinks are flaws, and then drops them for a newer face. Something like that happened with my initial hypothesis. Through the late 1960s, many economists were infatuated with it. They spent months poring over the data, manipulating them on the new computers that were starting to appear across academia, and eagerly submitting research papers to the academic journals. But 1972 proved to be a key year. By then, a new wave was sweeping through finance. Markowitz portfolio theory, Sharpe asset-pricing, and the Bachelier market model were spreading; and the next year Black and Scholes published their influential options-pricing formulae. "Modern finance" was the official religion. My hypothesis contradicted it; and I was about as welcome in the established church of economics as a heretical Arian at the Council of Nicene. In 1972, a University of Chicago grad student, Robert R. Officer, crystallized many of the qualms of the economics establishment. His Ph.D. thesis found evidence both for—and against— strict scaling in the same set of data. Other seemingly contradictory reports appeared. But the critics could not simply explain away the evidence supporting it; and I preferred just to forget about them. I found many of the criticisms were based on statistical tests inappropriate to the data—often a problem in statistics. But at that time, no one could devise a theory to reconcile all the conflicting data. That was still to come a bit later, and was not developed until "modern finance" had begun to falter.

The Meaning of Cotton

But why was the economics establishment so alarmed by the cotton research in the first place?

Remember the standard models. If the cotton price-changes fit the standard theory, they would be like sand grains in a heap; somewhat different sizes, but all sand grains, nonetheless. My cotton research showed something different: The changes were more like a mixture of sand, pebbles, rocks, and boulders. Some days, cotton prices hardly budged from the previous close; those are the small sand grains. Other days, the prices leaped a few percentage points; those are the boulders. Some days, there was no news in the market: quiet prices, sand grains again. Other days—perhaps word of a drought in Missouri finally reached New York—the news was big: wild price moves, statistical boulders. Together, all this news big and small, all these price-changes big and small, mix together in the crucible of a marketplace.

Fine, you may say. That explains the "fat tails," or abnormally big changes, in the cotton prices. But over a few years of daily trading, or a century of monthly trading, the same pattern emerges that you can see, with your own eyes, on a simple sheet of paper. Why the scaling? What does it mean that the price-changes scale?

Here, I can only speculate. In the physics I learned as a student, there is a clear barrier between the very large and the very small. At the very large scale of the cosmos, the relativistic space-time laws of Einstein apply. In the medium-scale world of our daily lives, Newtonian mechanics holds. And in the subatomic world of electrons and quarks, the entirely different laws of quantum mechanics apply. Three different regimes, three different scales, each one distinct from the last. The laws of physics do not scale. Shortly after my cotton papers (and quite independently), statistical physics expanded to study phenomena called "critical." It ignored and contradicted Zipf's distinction between physical and social science by discovering scaling relations of its own and fully explained them on the basis of unquestioned mathematical properties of matter. But economics is different. It lacks unquestioned mathematical laws to rely upon. Also, time, not space, is the scaling factor. Some time-spans matter, of course. In cotton, the annual cycle of planting and harvesting has a regular, periodic effect on trading; cotton stocks rise at harvest and trend downward until the next harvest. It is predictable. Economists routinely factor it out of their long-term analyses. But once the data are seasonally adjusted, is there any other time-scale that has a direct impact on cotton prices? After seasonal adjustment, is there anything in economics like the gulf between the quantum and Newtonian worlds? Do three weeks of trading really happen on a different economic planet than three days of trading, or three hours? Clearly not. All charts look the same.

Gaussian or not, scaling or not: Does it matter? Yes. First, it shows that prices can and do gyrate wildly. The market is very risky—far more risky than if you blithely assume that prices meander around a polite Gaussian average. Economists have long debated two opposing pictures of a commodity market. One views it as an insurance exchange, a financial machine for farmers and consumers to reduce their opposing risks, with the help of speculators as middlemen. Another views it as a wild casino, more risky than the stock market at its worst; while motives may differ from farmer to speculator, they are both gambling. The price data do not resolve that debate; insurance markets can be risky, too. But the data do help explain why commodity investing has limited appeal. Instinctively, most people regard a cotton contract as a riskier proposition than a Blue Chip stock—despite the fact that, by the standard analysis, commodity investments should play a bigger role in the portfolios of the wealthy. Most people sense the greater risk, and shun it. Perhaps no great statistical analysis was needed at all:

This fact of mass psychology, alone, might have been sufficient evidence to suggest there is something amiss with the standard financial models.

Second, as will be seen in the next chapter, data that scale can produce surprising patterns—patterns that, if you glance at them, you would swear are periodic, predictable, and bankable. Anyone studying the cotton price records could easily imagine he was seeing "corrections," "resistance levels," and the other signals that a technical analyst seeks to buy, sell, or hold. It is fool's gold.

Lastly, the cotton story shows the strange liaison among different branches of the economy, and between economics and nature. That cotton prices should vary the way income does; that income variations should look like Swedish fire-insurance claims; that these, in turn, are in the same mathematical family as formulae describing the way we speak, or how earthquakes happen—this is, truly, the greatest mystery of all.

Let us mull the promises that science makes to society to win its support. The grand promise is to endeavor solving the great mysteries—to the list of which I have added one. But there is also a more practical promise. It consists in helping society to improve, to prevent it from acting on the basis of theories that sound nice but are not true to the facts, and to help it act on the basis of facts—even if those facts have yet to find a theory that fully explains them.

Coda: Looney 'Toons, Reprised for Long Tails

Pictures aid understanding; hence, my frequent use of schematic diagrams or cartoons. The chapter on turbulence showed how a simple fractal process can generate a complex imitation of a financial chart according to the Bachelier model. As a final thought here, I show how the ideas of scaling and discontinuity can be translated into such a cartoon. The aim: To show the subtle link between the "fat tails" and abrupt price-changes of real financial charts and the abstractions of fractal analysis.

A recap: We started the Brownian cartoon with a rising, straight-line initiator, and a zigzag generator. We made copies of the generator, shrank them, and interpolated them into the diagram so that

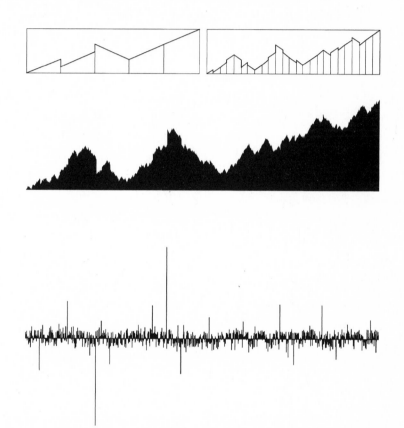

A cartoon of discontinuity. There are many ways to illustrate the crucial concepts of fat tails and discontinuity—and this one employs the kind of fractal process used earlier in this book. In contrast to the Brownian cartoon shown previously, this diagram has a much more complicated fractal generator. It begins with five inclined intervals, but adds two vertical discontinuities. The second panel shows the completed construction, and the bottom panel shows the changes in value from one moment to the next. The outcome: A chart that exhibits a species of scaling and wild variation akin to that of cotton prices.

everywhere a straight line appeared it was replaced with a zigzag pattern to fit. By repeating this process over and over, a jagged, complex chart gradually appeared. By careful design, the specific kind of chart shown before was of a Brownian motion—the standard model underlying conventional financial theory. What made it so was the specific shape of the generator: Starting at the point (0, 0), it rose to the point (4/9, 2/3), fell to the point (5/9, 1/3), and ended up at (1, 1). A key observation regards the size of the three segments of the generator. Their widths were 4/9, 1/9, 4/9. The heights: 2/3, -1/3 (minus, because the line falls), and 2/3. Look closely at those six numbers. Each width is the square of each height. It is a nice, tidy relationship—just the kind of thing you would expect from a well-mannered Brownian motion.

Change those coordinates, and almost invariably something very different emerges. In particular, the outcome can be much more like what I observed in the cotton price charts. As shown below, start with a generator broken into three equally spaced widths: each one, a third of the box wide. In each, there is a positively sloped line that rises one-half the box's height. But then something is added: two vertical jumps, the first up one-half a unit and the second down by a full unit. Unlike the Brownian generator, this one has sharp discontinuities. Each stage of interpolation automatically adds further jumps endowed with a special kind of order: The positive and negative jumps, taken separately or together, follow a power-law distribution. Scaling has generated "fat tails," which can be measured by an α exponent—just as in the Pareto or L-stable cases discussed earlier in this chapter. Playing with the generator one can "tune" α and the degree of asymmetry between the tails.

To what end is all this? To show, through scaling and fractals, the strange nexus between two seemingly disparate cases—one the familiar randomness of a game of tossing coins, the other the complexity and risk of a cotton price chart.

CHAPTER IX

Long Memory, from the Nile to the Marketplace

IN 1906, a young Englishman named Harold Edwin Hurst arrived in Cairo. It was to have been a short stay. But it lasted sixty-two years and ended with his solving one of the great mysteries of the pharaohs—and, inadvertently, providing a clue to the way financial markets work.

Hurst's problem concerned the Nile floods. The great river's waters, wrote Greek traveler-historian Herodotus about 450 B.C., start rising at the summer solstice, mount to a crest over the next one hundred days, and carry such rich soil downriver that "the people get their harvests with less labor than anyone else in the world." What caused the floods? The oldest legends had the Nile flowing from the full breasts of the goddess Hapi. The priests said it fell from the heavenly cataracts of the Celestial Nile. Some said northerly winds blowing in from the Mediterranean backed the river up. Others said the floods came from melting snows far to the south—a yarn Herodotus dismissed; surely no snow could fall in a southern heat so intense it even burned the people black, he reasoned. "About why the Nile behaves precisely as it does I could get no information from the priests or anyone else."

More directly important than *why* the Nile flooded, however, was *how much* it flooded from one year to the next. Sometimes the floods ran high, sometimes low. Prosperity or poverty hung upon the outcome. Recall the Bible story of the seven fat years and seven lean years, foretold from Pharaoh's dream by Joseph, son of Jacob. By Hurst's day, when the land of Egypt had passed to British rule, the problem was no less pressing. The population in the Nile Valley was swelling; Manchester mills wanted Egypt's cotton; and the river's dams were unequal to the task of managing so vast and precious a resource. Hurst, who rose to become chief scientist of the public-works ministry, was charged with devising what was called "century storage" to stockpile water against the worst possible droughts. It was a daunting task. He observed: "There has been a great deal of investigation as to trying to forecast the flood; nothing of any practical use has come out of it, so you don't even know one year what the next year's flood will be like."

But he did find a formula for the floods—and his work earned him the respectful nickname, Abu Nil, or Father Nile. What is more, he found that the same formula also applied to a broad range of other phenomena: the way clay layers accumulate on a Crimean lake bed, the annual pattern of rainfall in New York, the growth of tree rings on Pike's Peak. That is not all. Picking up Hurst's trail in the 1960s, I discovered the same "Nile pattern" in many other contexts—oddest of all, in how a stock price fluctuates. The Nile pattern is a crucial part of fractal geometry. Other researchers, broadening my work in recent years, have found it in international crude oil prices, London gold fixings, and the deregulated U.S. electricity market. Indeed, the Nile pattern provides the second major link in my theory of how financial markets work—a necessary complement to the first one about scaling laws and "fat tails" of the last chapter. Adding a third link in a later chapter, a comprehensive market model results.

Good scientists are often circumspect in their formal, academic-journal utterances. The most famous understatement in science may be Watson and Crick's one-sentence observation, in their original 1953 *Nature* report on the structure of DNA, that "it has not escaped our notice that the specific pairing we have postulated immediately suggests a possible copying mechanism for the genetic material." Hurst was similarly cautious in the first report of his Nile discovery, in the dense pages of the 1951 *Transactions of the American Society of Civil Engineers*. "It is thought that the general theory," he wrote, "may have other applications than the design of reservoirs for the storage of water."

Abu Nil

H.E. Hurst was a model civil servant of Imperial Britain. He was born in 1880, the son of a village builder of limited means whose family had lived near Leicester for almost three centuries. He left school at fifteen, trained in chemistry and carpentry, and after evening classes won a scholarship to Oxford at twenty. To everyone's surprise—as he later described it to me—he won a first-class honors degree in physics, despite a lack of early preparation in mathematics. But it was in Egypt that he found his future.

As the twentieth century began, the British Empire had finally put down the fundamentalist Mahdi revolt upriver in Sudan. A period of relative peace, growth, and dam construction ensued. For most of its northward course the Nile was undisputed property of the British Empire: from Lake Victoria to Lake Albert, to the joining of the White Nile and the Blue Nile at Khartoum, over the swamps, clay, and cliff-lined basins of Sudan and southern Egypt, and out at last to the broad Delta on the Mediterranean Sea. For even as vast an empire as Britain's, the Nile's scale was immense. The river was 4,160 miles long. Its broad basin covered 10 percent of the land area of the entire African continent. Its waters were bounteous. Its average annual discharge over a century was 92.4 billion cubic meters—enough that, Hurst observed, if even an eighth of that sum were transported miraculously to Yorkshire it would flood the county in two and a half feet of water.

From the completion of the first big dam at Aswan in 1902, British technology and industry were fully deployed to exploit the Nile's economic power, control its floods, and expand the irrigable land. When Hurst arrived, his first task in Cairo was quaint: He transmitted the official time from the observatory to the citadel, for the firing of a midday gun. But with his scientific training, he was soon drawn into the great Imperial project of mapping and measuring the river. He traveled by boat, by foot with porters, by bicycle, by car, and later by plane. British engineers and their Egyptian helpers deployed current meters to count the revolutions per minute as the water rushed past. With lead weights, piano wire, and trigonometry, they sounded the river's depths. They built new flood-level gauges of marble set in masonry. In the Sudan, where, Hurst reported, "the topsoil is often a gray clay called cotton soil which shrinks and cracks in the dry season and swells and rises in the rains," they sank screw-piles deep down into the permanent subsoil to anchor the flood gauges. They measured the swept-along sand, clay, and silt, observing murky concentrations peak in late August before the flood's crest and fall to clear water in the winter. They ventured with their current meters into the great swamps along 450 miles of the river through Sudan—a region generally avoided by conquerors since the time of Nero's centurions. And where the explorers Stanley, Speake, and Burton only a few generations before had tread with difficulty and trepidation, the new generation of pragmatic British surveyors journeyed with transits, levels, and slide rules to map the uncharted tributaries on a scale of one to fifty thousand.

Their primary goal was regulating the river. For hydraulic engineers of Hurst's day, a river's routine season-to-season fluctuations were well understood. But the year-to-year variation on so vast a river was an entirely different problem. The Nile discharges ranged wildly, from 151 billion cubic meters in the wet year of 1878–1879, to 42 billion cubic meters in the drought of 1913–1914. Moreover, that dry spell was followed only two years later by another. Wet years clustered together, too. Yet there was, Hurst wrote, "no obvious periodicity." How can you control something with no predictable pattern?

The obvious solution was a high dam—high enough to hold back the waters of several wet years, and release it during a run of dry years. But how high is that? Dam design was an important task in the nineteenth century, but one in which—like finance today—the mathematically easy path was preferred. Engineers assumed flood variations from one year to the next were statistically independent, as with Bachelier's coin-tossing. With coins you can get runs of heads or tails, of course; otherwise, there would be no winner. And there is a simple formula for it: The range between Harry's biggest winnings at one moment of the game and his worst loss at another time varies by the square root of the number of tosses. For instance, say the game lasts 100 tosses, and Harry's biggest gain was eight and his worst loss was three. The range from best to worst score was eleven. Now say the game goes on 100 times longer, for 10,000 tosses. The formula says the range should be about ten times greater, or 110. Now, the theory suggests, Harry's best score might be sixty-seven and his worst, minus forty-three. So, taking the cue from Harry, a hydraulic engineer can make a few simple calculations. Say he wants to replace a twenty-five-year-old dam with a higher one, proof against one hundred years of flood. The time-scale of the new dam is four times that of the old. So, if conventional math applies, the new dam should be twice as high as the old. Tidy and simple.

But also wrong. In fact, the dam should be higher than that, Hurst concluded. He found the range from highest Nile flood to lowest widened faster than the coin-tossing rule predicted. The highs were higher; the lows, lower. But the problem was not the individual floods; looked at singly, the bell curve fit the data on each year's flooding reasonably well. Apparently, it was the *runs* of weather—the back-to-back floods or droughts—that were changing the game. It seems obvious, now: Not just the size of the floods, but also their precise sequence, matters.

Hurst, studying the flood records, devised his own formula to capture this effect. To do so, he began with the Nile, then looked far beyond, without any preconceptions. He gathered records of discharges from Lake Huron and the Truckee River near Lake Tahoe. He looked at the annual water levels in Sweden's Dalalven Lake; rainfall measurements from Adelaide, Australia, to Washington, D.C.; the thickness of lakebed sediments in Russia, Norway, and Canada; temperature readings from St. Louis to Helsinki; the pattern of tree rings in Flagstaff pines and Sequoia—even sunspot numbers. He looked through any reliable, long-running records he could find that were in any way related to climate, for a total of fifty-one different phenomena, 5,915 yearly measurements. In almost all cases, when he plotted the number of years measured against the high-to-low range of each record, he found the range widened too quickly—just like the Nile. In fact, he found as he looked around the world, it all fit the same neat formula: The range widened, not by a square-root law as in coin-tossing, but as a three-fourths-power (0.73, to be precise.) A strange number; but it was, Hurst asserted, a fundamental fact of nature.

Hydrologists were skeptical. From 1951 to 1956—Hurst was then in his seventies—he published three lengthy essays reporting his findings. Each one was accompanied by a flurry of printed comments from supporters and detractors. Some praised him, finding more records to support his case. Others accused him of statistical voodoo. One, a Mr. F. A. Sharman, senior civil engineer at Sir William Halcrow & Partners, sarcastically observed that anyone who claimed to find a common thread from tree rings to sun spots to mud layers must have taken "a sensational step towards finding

ena, why not in hydrological cycles, too? Following this particular lead should be easy. A few hundred yards from my lecture hall were the offices of the Harvard Water Resources Center, and a leading hydrologist, Professor Harold A. Thomas Jr. An improbable site, ivy-clad Harvard, for a gritty school of dam-building. But Thomas quickly told me about Hurst's observations. I thought I had it figured out in a flash. Hurst's law, that the range from high to low flood levels widened by the three-quarters-power of the standard deviation, sounded like a mere variant of my cotton formula. The big floods, I reckoned, were like big price jumps; the disastrous droughts were market crashes.

But great theories are often humbled by mere facts. It was not that simple. Looking up Hurst's papers revealed that his point had not been the size of the variations, but the precise sequence of them. If jumbled up and taken out of their original sequence, his data yielded nothing special at all: a boring bell curve. Now I was hooked. When studying cotton, there had been obvious correlations between past and future prices; I mentioned this at the time, but I could not develop it further. Therefore I had pushed the precise sequence of prices aside for later study, working as if each had been independent from the last. Hurst's research posed yet another mystery. As a valuable added charm, it was a truly ancient one, as old as the pyramids.

How much does the past shape the future? A moral philosopher would phrase it this way: Is it fate that determines our course, or do we choose our paths afresh with each new decision? A mathematician trades in another terminology: Is one event dependent on another, or independent from it? If Event B is dependent on Event A, then A's occurrence changes the odds of B happening. If a basketball player sinks two shots in a row, evidence suggests, odds are greater that his third shot will also score. By prowess or psychology, a player can have "hot" streaks; successive shots are, to some degree, dependent on one another. But how long will his scoring streak last? Is it broken after just one miss? Two? Five? Over how many

shots, precisely, does the "hot-hands" effect linger? Put in mathematical terms, over how many time-periods is the dependence significant? Now look at it from another perspective. Suppose you are
a spectator watching the game from the stands. How many misses
would it take before you conclude the player is no longer hot?
Three? Seven? What looks dependent at first glance is not necessarily so on closer study. As any chartist has learned to his sorrow, the
most random and independent events can spontaneously appear to
form patterns and cycles.

Economists think about this in reductive fashion. First, as I have
said, most of their financial models assume—incorrectly—that one
day's price is independent of the last; it takes a random walk. But
with economic quantities—production, inflation, unemployment—
some form of dependence is the rule, and economists crank the
numbers through cookbook tests to measure how strong it is, and
over how many time-periods it extends. If inflation jumps up in
April, how likely is it to rise in May? How about two periods later,
in June? Three? For each time-lag, economists measure the
strength of the correlation, and that strength can vary between an
arbitrary value of 1 for events that move in perfect lockstep, and −1
for events that always zig when the other zags. Zero, in the middle,
means no correlation at all; events bounce around with no regard to
one another. There are an infinite number of intermediate values on
that −1 to +1 correlation scale. Each one tells a different story of the
sign and strength of the short-term dependence. Most often, the
strongest correlations are the short-term ones between periods close
together; the weakest are those between periods far apart. If you
plot all the correlations, from short-term to long-range, you get a
rapidly falling curve. How fast it falls varies from one economic
quantity to another. Inflation is "persistent": Its curve falls rather
slowly. Once inflation gets going, it is difficult to slow—as central
bankers discovered in the 1970s. The curves of many other economic quantities have peculiar bumps on the way down. Corn
stocks are like that; the curve bumps up at the one-year mark,

because the annual cycle of sowing and harvest has a powerful effect on supply. Gross domestic product, the standard measure of an economy's output, has several odd bumps on the way downhill to zero correlation—often, at a few years, at fifteen to twenty years, and at forty to sixty years. Economists have been debating the bumps' meaning in the business cycle for years, with no clear answers.

But why stop at fifteen or fifty years? Hurst's work suggested something more radical to me: correlations that decrease, but so slowly that they seem never to vanish completely, no matter how far back in time you go. How is that possible? Recall that Hurst was ultimately interested in reservoir levels; his range formula is mathematical shorthand for calculating the optimum dam height and reservoir level. Say a series of wet years fills the reservoir. Then, some years of mostly moderate weather follow—but the reservoir is full; the prior wet years are still having an effect. Then some dry years arrive. Now the reservoir is emptying. But it has more water than it otherwise would; still, the prior wet years are having an effect. You can get a glimpse of this in the chart below, of ring-widths in some of the world's oldest trees, ancient bristlecones on Mount Campito, in the White Mountains of California. The curve starts out as in most such charts, called correlograms, with high correlations for short time-periods: Adjacent tree-rings, the marks of growth only a year or two apart, are highly correlated. Beyond a few years, the correlations fall; the pattern from one decade or century to the next is more haphazard. But the correlations fall more slowly than expected. In fact, it is 150 years before they are so insignificant that to distinguish them statistically from chance, the usual tests are powerless. I had to devise new ones inspired by Hurst. And why are the rings correlated over so many years? There lies a global warming debate.

This is long-term dependence. It is a subtle concept; so let us begin with an example that gives the general idea. A pure radioactive substance decays geometrically in time. After one half-life, only

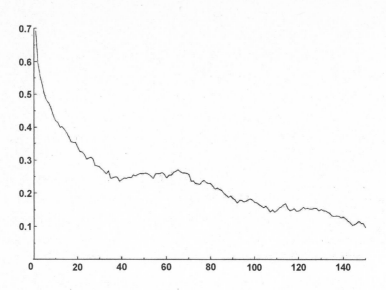

Long memory.   Start with an event—say, a cold year that slows tree growth. Will the corresponding tree-ring the next year be as narrow as the first, or wider? What about a decade later, or one hundred years? This chart, from Baillie 1996, is called a correlogram. It shows how the correlations in tree-ring sizes at Mount Campito, California, change from one time-period to another. It is a long, slow decline—slower than one would normally expect.

half is left; after two half-lives, only a quarter; then an eighth; and then it is practically gone. But consider a mixture of different radioactive substances, such that very short, medium, long, and very long values of the half-life are present. When the short half-life components are practically all gone, the others have barely begun to decay; their effect will endure. That is long-term dependence. Nor is this a hypothetical example: It is a fact and—last time I looked—a real mystery that the garbage from nuclear explosives is a radioactive goulash with a large number of values for the half-life. In nearly all other cases, the mixture idea is just a metaphor, but it helped me conceive of long dependence as a way to account for Hurst's findings. It is a pillar of fractal geometry.

Now think of finance. In 1982 IBM, then the world's biggest computer company, decided some upstarts at Apple were threatening its future with a new product called the personal computer. Uncharacteristically, IBM acted quickly. It bypassed its own big chip factories and software departments. It picked a struggling semiconductor company named Intel to make its microprocessors and a bright but insignificant kid named Bill Gates to provide its software. The rest is well-known: Intel and Microsoft grew wildly, beyond any imaginable bounds. IBM stumbled, and shrank. But the fates of these three companies are still intertwined. Their stock prices affect one another, as profits or troubles at one redounds on the business or market-ratings of the others. That event of three decades ago, IBM's midwifery to two new industry giants, continues to reverberate today in IBM's stock price. The dependence there is about thirty years long. One can easily imagine even longer dependence: The court-ordered breakup of John D. Rockefeller's Standard Oil Trust in 1911 continues to affect its surviving children today, ExxonMobil, ConocoPhillips, ChevronTexaco, and BP Amoco.

No one is alone in this world. No act is without consequences for others. It is a tenet of chaos theory that, in dynamical systems, the outcome of any process is sensitive to its starting point—or, in the famous cliché, the flap of a butterfly's wings in the Amazon can cause a tornado in Texas. I do not assert markets are chaotic, though my fractal geometry is one of the primary mathematical tools of "chaology." But clearly, the global economy is an unfathomably complicated machine. To all the complexity of the physical world of weather, crops, ores, and factories, you add the psychological complexity of men acting on their fleeting expectations of what may or may not happen—sheer phantasms. Companies and stock prices, trade flows and currency rates, crop yields and commodity futures—all are inter-related to one degree or another, in ways we have barely begun to understand. In such a world, it is common sense that events in the distant past continue to echo in the present.

In the 1960s, some old-timers on Wall Street—the men who remembered the trauma of the 1929 Crash and the Great Depression—gave me a warning: "When we fade from this business, something will be lost. That is the memory of 1929." Because of that personal recollection, they said, they acted with more caution than they otherwise might. Collectively, their generation provided an in-built brake on the wildest forms of speculation, an insurance policy against financial excess and consequent catastrophe. Their memories provided a practical form of long-term dependence in the financial markets. Is it any wonder that in 1987, when most of those men were gone and their wisdom forgotten, the market encountered its first crash in nearly sixty years? Or that, two decades later, we would see the biggest bull market, and the worst bear market, in generations? Yet standard financial theory holds that, in modeling markets, all that matters is today's news and the expectation of tomorrow's news.

A Random Run

A nice idea, long memory. But what do you do with it?

Go back to the original Brownian motion, of individual particles in water. How far will a molecule get from its starting position in two nanoseconds, or two hours? The square-root rule, mentioned earlier, applies: A Brownian particle that travels one hundred seconds will get around ten times farther than one that travels just one second. When applied to prices, under the standard financial models, this is all very handy. It tells you how far, in any given holding period, an asset's price may rise or fall and how much it is likely to fluctuate within that broad band. Brownian motion is a bank economist's best friend. When asked by his boss to predict the dollar-sterling rate a year from now, he can smartly sidestep the question. Working from today's rate of $1.65 to the pound, he gives, not a specific forecast of $1.70, but a vague Brownian range: "The pound will trade between $1.55 and $1.75, and there is a good case to make for it trending up within that range—if the U.S. economy stutters, if inflation in Britain rises moderately, if . . ." Of course, he is only staking his job on the vague range, not on the what-ifs—so he survives to forecast another year.

But what happens if the exchange rate wanders farther than the square-root-of-time law forecasts? Trouble for the economist, obviously. How could that happen? Easily, if exchange rates exhibit long dependence. A rate move in one direction will tend to continue on the next day, and a few days later. The rate will still bounce around from day to day. But in the long run it will drift farther and farther from its starting point. The rates are no longer fluctuating by the blindest form of chance. Now the game is rigged, just like the Nile floods running in sequence.

In fact, you can put a number on this tendency towards cheating: I call it H in honor of Hurst, but also out of respect for an earlier, very pure mathematician, Ludwig Otto Hölder. (Oddly enough, just for fun he had been dealing with similar thoughts.) The formula starts like the familiar Brownian case: The distance traveled is proportional to some power of the time elapsed. But now, the power is no longer a square root, or one half. It could be any fraction between zero and one, and each produces a totally different type of price series. If H is bigger than the Brownian 0.5—say 0.9—the price will roam far; its motion will be "persistent," like a mule intent on heading in its own direction no matter what the rider does. Of course, it will eventually reverse: Overall, the increments still have to fit the bell curve. So for every "plus" direction, there is a "minus" direction, but they can cluster together, as with the Nile floods. Now the opposite case: If H is smaller than the Brownian 0.5, say 0.1, the price or particle will roam less. Each step will tend to be followed by another reversing direction, and then another back the other way, and so on in a narrow and furious zigzag pattern. It behaves like a frightened horse that prefers to stick to the safe path rather than obey its rider and gallop off into the dark fields to left or right.

The diagrams following illustrate the point. They show, not the position of the Brownian motion, but the changes or steps up or

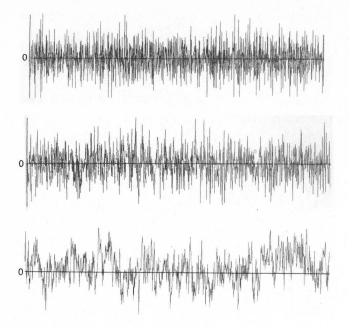

Spot the trends. The standard financial models assume each price change is independent of the last. And if that is wrong? These diagrams, drawn by a Calcomp pen tracer, are antiques of computer graphics. They represent three models of price increments. The middle chart shows the standard financial model, which assumes each price change to be independent of the last. In this case, the parameter H that measures the dependence equals 0.5. The bottom chart shows the case when prices have a tendency to keep going in the same direction; in other words, price changes can run in long streaks of positive values—or, conversely, negative. Here, $H = 0.9$. The top chart shows the opposite phenomenon: once pointed one way, the motion will tend to reverse and head the other way. Here, $H = 0.1$.

down from one instant to the next as time proceeds. The bottom one shows the persistent case, when H is big and the resulting price trends are broad. The middle shows the classical Brownian case. And the top shows the "anti-persistent" case, when H is small and the action is furious but still constrained. Because of the fractional values that H can adopt, I denoted the sums of those interdependent increments as "fractional Brownian motions."

It is a peculiar property of most long-memory processes that *seeming* patterns arise and fall, appear and disappear. They could vanish at any instant. They have no real permanence. They cannot be predicted. Look again at the persistent fractional Brownian motion charts, the top and bottom ones. You can spot intervals in which the motion appears to trend upwards, or slide downwards. Mere chance, of course. If you tried to bet on such a trend, you might win money for a while; but you could as easily lose it if your timing was wrong. Pictures can deceive as well as instruct. The brain highlights what it imagines as patterns; it disregards contradictory information. Human nature yearns to see order and hierarchy in the world. It will invent it where it cannot find it.

The Selling of *H*

New ideas often struggle to make their way in the world. I wrote three essays in 1964 and 1965 as my theories of dependence and scaling evolved. They drew disappointingly little notice. They ran counter to conventional wisdom, and were unwelcome at established journals. Some friends suggested my problem was style, not substance. The highest priority, they convinced me, was to bring out a simple and clear mathematical exposition. So I joined forces with a young mainstream mathematician, John Van Ness, then at the University of Washington in Seattle, in hopes of producing a more orthodox paper. Alas it, too, was soundly rejected, on the grounds it had nothing new to say (I may have been to blame for that, as I had insisted we amply cite any prior mathematician and economist who had come within a mile of our ideas). In the end, two long years after we had written the paper, a chance meeting with a journal editor at a dinner party found it a publisher in 1968—albeit an obscure one: The *SIAM Review* of the Society for Industrial and Applied Mathematics. If the time a modern scientist must lavish on publicity were redirected to discovery, what marvels would we see?

In the meantime, I resolved to show how the theory could be applied. So I began modeling water storage, just as Hurst had done. I tried collaborating with a Harvard hydrologist, but his computer program produced garbage. (For which he blamed me: The course of scientific collaboration seldom runs smoothly.) Then chance again intervened. In the fall of 1967 IBM hired a former government hydrologist and Harvard post-doc named James R. Wallis. IBM was trying both to look more "earth-friendly" in keeping with the times and to expand its research into ecological computer applications, including river networks. Wallis and I worked well together. When it came time to publish, we took no chances: We went directly to the editor of the leading hydrology journal, *Water Resources Research.* To persuade him to publish, we prepared elaborate computer graphics of my water model, produced laboriously on an excruciatingly slow Calcomp tracing-pen plotter. It was state-of-the-art in 1968, but its output was so faint that for later publication the graphics had to be retraced by hand on vellum, and then photographed.

The journal editor, Dr. Walter Langbein, was a high-ranking official in the U.S. Geological Survey—and a dry-looking, gray gentleman. A tough customer. We met him and some fellow officials during a meeting of the Geophysical Union in Baltimore. In a hotel room there, we unrolled our printouts one by one before him. The game: Could he distinguish the graphics of real hydrological data from the fake ones? And of the forgeries, could he tell which were based on my fractional Brownian motion calculations, and which were drawn using the conventional hydrology models? The latter were, for all practical purposes, identical to those of the "modern" theory of finance. He paged through the mess spread on his bed, and immediately spotted some fakes—including one based on his own research. He laughed. He knew it had been a crude model, he said, but he had not realized how crude. After some more sorting, he gave up. He could not tell our models from the real data. We showed him the key, written on the back of each illustration—and by then he was more than willing to listen to how our model worked. He agreed on the spot to publish, without the usual routine of calling academic referees. Such forceful editors are the salt of the earth, but rare in scientific publishing. More common is the risk-avoiding bureaucrat, nailed to an influential editorial chair.

In economics, acceptance was harder and accompanied by many misunderstandings that persist today. Economists were talking about dependence, of course. In 1965, Irma Adelman, then an associate professor at Johns Hopkins, wrote an essay, "Long cycles—Fact or artifact?" The next year, Clive W. J. Granger, a young mathematician from the University of Nottingham, England (who in 2003 won a Nobel), moved from question to assertion: The "typical" economic variable, he wrote, has very long-term correlations. Long-range, yes; but infinite memory, as I was asserting? *Vade retro Satanas!* Go back, Satan! In the end, economists began finding evidence—in gold prices, in oil markets, in foreign exchange. But they also found plenty of markets that did not fit the theory. My own research showed that prices for cotton, wheat, and British government bonds behaved independently. In fact, something odd was dropping out as the facts were sifted: The degree of dependence varied significantly from one type of financial asset to another. That degree appeared to be captured by H, my measure of how far a random run would go. Could H be a new yardstick for finance, like the Dow, beta, or other numbers loved by Wall Street?

I set to work again, this time with Murad S. Taqqu, whose Columbia University Ph.D. thesis I was supervising. The young statistician was also my paid research assistant and rewrote a computer program for testing fractality and estimating H. Again, this was in the days of expensive computers and difficult programming languages. The only time we could stay for awhile on IBM's biggest "iron" was over a long Christmas weekend. It ate through reams of prices we fed it, and yielded a boxful of Calcomp outputs. Such research confirmed a puzzlingly intricate range for H. Interest rates on loans from banks to brokers, "call money" in Wall

Street parlance, were fairly dependent: Its H was 0.7, meaning it tended to move up and down in long, persistent trends—perhaps because it was just following similarly broad trends in the economy. Wheat and U.K. bonds were about 0.5: independent, as assumed in the standard financial models and in my 1963 model for cotton.

To be frank, the pattern is not yet clear. Theories abound. Some speculate that a high H is what you would find in a very risky "momentum" play, where emotional crowd-behavior can more easily sweep investors along. By contrast, an H closer to 0.5 would imply a very random, heavily arbitraged market—more compatible with the classical Brownian model of how markets should work. For instance, Edgar E. Peters, chief investment officer of a Boston fund manager, PanAgora Asset Management, reported finding high H's of 0.75 for Apple, 0.73 for Xerox, and 0.72 for IBM. More boring stocks had lower values: Anheuser-Busch, 0.64, Texas State Utilities, 0.54. In the foreign exchange market, some economists have found currencies closely tied to the U.S. dollar, such as the Canadian dollar, tend to have near-Brownian H of 0.5. Others, such as the Malaysian ringgit, look more like high-H technology stocks. Such research requires great caution, however. The quality of data, the care in analysis, and even the fundamental methods employed can vary from one study to another. In 1991, an MIT economist, Andrew W. Lo, published a weighty rebuttal of my claims for H. He reported that my statistical tests could confound long-term memory with the effects of short-term memory. But shortly after, other economists said his tests, in turn, were potentially flawed. A lesson arises from this: Never hurry *and never publish* any result based on a single tool.

Besides, the whole field turns out to be more intricate than any one simplistic test could resolve. A long dependence fully described by a single H is a very special case, that of the fractional Brownian motions shown in the charts on page 188. It is also possible to have a multitude of distinct H-like exponents. For instance, in dollar-

Deutschemark exchange rates, one of those exponents suggests the price changes are independent of one another; but the other exponents say they are dependent—and it is the latter that are right. A complicated situation it is, indeed, nothing like the nice, simple coin-tossing model of "modern" financial theory.

But why all the fuss?

The whole edifice of modern financial theory is, as described earlier, founded on a few simplifying assumptions. It presumes that *homo economicus* is rational and self-interested. Wrong, suggests the experience of the irrational, mob-psychology bubble and burst of the 1990s. A further assumption: that price variations follow the bell curve. Wrong, suggests the by-now widely accepted research of me and many others since the 1960s. And now the next assumption wobbles: that price variations are what statisticians call i.i.d., independently and identically distributed—like the coin game with each toss unaffected by the last. Evidence for short-term dependence has already been mounting. And now comes the increasingly accepted but still confusing evidence of long-term dependence.

Some economists, when thinking about long memory, are concerned that it undercuts the Efficient Market Hypothesis that prices fully reflect all relevant information; that the random walk is the best metaphor to describe such markets; and that you cannot beat such an unpredictable market. Well, the Efficient Market Hypothesis is no more than that, a hypothesis. Many a grand theory has died under the onslaught of real data.

Coda: Looney 'Toons of Long Dependence

As an aid to understanding, it is cartoon time again. As shown in prior chapters, fractal geometry allows for synthesis that starts from some simple ideas and generates complex structures. We began with an approximate Brownian-motion diagram, and later obtained a

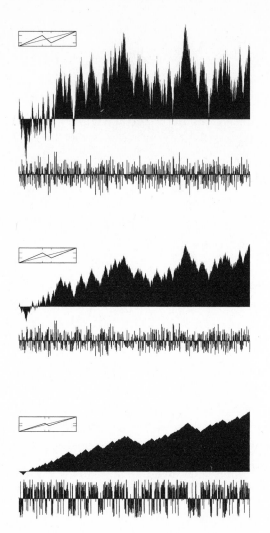

A cartoon of long dependence.   One can illustrate the concept of long-term dependence with fractal cartoons of the sort used earlier in the book. Here, the middle panel shows the standard model of independent price changes; H, the index of dependence, is 0.5. The small inset box shows the fractal generator; the black fever chart is the completed, randomized construction; and the bottom line shows the "price" changes. The bottom panel corresponds to H greater than 0.5. Its behavior is called persistent, as can be seen most clearly in the difference chart just below it: It has runs of positive values that persist for a while, before switching to negative runs that, in turn, also persist. The top panel shows what happens when H is less than 0.5. Its "anti-persistence"—the exact opposite behavior—is best seen on the wildly fluctuating fever chart.

fat-tailed, discontinuous price chart. Now, the same process can be used to illustrate the theme of the current chapter, long-term dependence.

In the preceding diagram, the fractal initiator is again a rising straight trend line, and the generator is a simple zigzag pointing up, down, and up again. In the Brownian case, we made the width of each interval the square root of its height—that is, the width equals the height to the power of one half. Call that power H. No coincidence there. As described earlier in this chapter, a Brownian motion has no dependence—each increment is unaffected by past or future changes—and the H exponent describing its behavior is exactly one half. So what happens if we change the exponent in the generator of the fractal diagram?

As shown, make each interval's height equal to its width raised to some arbitrary power between 0 and 1. A value of H greater than one half is shown on the bottom panel: It generates a final fractal diagram that shows persistence. Thinking in terms of the Nile, a record of floods that shows a sequence of mostly wet periods interspersed with brief droughts—or the opposite. Thinking in terms of prices, a long sequence of periods of growth with brief downswings—or the opposite. A value of H smaller than one half, shown on the top panel, has strong "anti-persistence": Successive changes tend to cancel each other out. Again, the power of fractals shows a strange connection among seemingly unrelated phenomena.

CHAPTER X

Noah, Joseph, and Market Bubbles

I will cause it to rain upon the earth forty days and forty nights; and every living substance that I have made will I destroy from off the face of the earth.

Genesis 7: 4.

What God is about to do he showeth unto Pharaoh. Behold, there come seven years of great plenty throughout all the land of Egypt: and there shall arise after them seven years of famine; and all the plenty shall be forgotten in the land of Egypt; and the famine shall consume the land.

Genesis 41: 28-30.

BEFORE WE CONTINUE, LET US RECAP. Models are important in science. They help us understand. If, on a computer, we can build a small-scale model of the global climate, of a planet's orbit, or of an economy's growth, we can test our knowledge. Models also help us act. In economics, to build a model is to fashion a tool. An econometrician who models a nation's current-account deficit is trying to glean some insight into future exchange rates. A financial analyst who designs a model portfolio is trying to test his investment hunches, and so advise clients. And a banker who models market volatility has one ultimate aim: to measure risk and, if possible, reduce it. On Wall Street, models are not toys; they are the high-tech arsenal on which millions are spent, by which one brokerage house or bank hopes to gain a slight edge over all the rest.

For the past forty years, my mathematical hobby has been making scale models of markets. Along the way to my main work, the development of fractal geometry, economic study, and financial models proved major milestones. I began with cotton in the early 1960s, as discussed earlier. But I moved on to railroad stocks, IBM shares, interest rates, currencies, and other assets. My trajectory, while not a random walk, has certainly been wide-ranging. With each new model, I tested new fractal properties, "fat tails," and long-range dependence. The next chapter will combine them through the notions of "trading time" and multifractals. But I am not yet finished; nor do I believe we are ever likely to have perfect understanding of so complex a system as the global money machine. In economics, there can never be a "theory of everything." But I believe each attempt comes closer to a proper understanding of how markets behave.

An Alien Plays the Market

In building a model, start simple. What are the few most important facts about price charts—the essential characteristics that, reduced to mathematical formulae and lines of software, would be the silicon heart of our computer simulation?

Step back, a long way. Imagine yourself a visitor from another planet, observing the millions of charts we scatter across our newspapers, magazines, televisions, and the Internet. They must be important to us, you deduce; their sheer volume suggests that. But what they mean, how they arise, why they look the way they do—all are mysteries.

So you begin by observing, carefully. Two obvious facts jump out: One, the prices bounce around a lot; and, two, they appear to move in irregular trends. In fact, you hear that some of these earth-dwellers try to gain power over one another by betting on these trends to amass wealth—but they usually lose.

The jumps can be quite large, indeed. Days of minor fluctuations, of less than a percent, can be punctuated by great leaps upward or falls downward—3 percent, 17 percent, even 40 percent in a day. That is wild variation: ungovernable and seemingly unpredictable spasms of movement. When you analyze it, you quickly see it does not fit the tidy pattern of the bell curve (known throughout the civilized universe as the epitome of mild, manageable variation). There are too many very big and very small changes, not enough medium-sized ones. And the changes appear to scale with time: The proportion of bigger to smaller price-moves follows a regular pattern as you look at monthly, weekly, or daily charts. In fact, if you consider only how much the charts wiggle, at different time-scales they all look roughly alike—and all very bumpy.

Now you look at the irregular trends. The size of the price changes clearly cluster together. Big changes often come together in rapid succession, like a fusillade of cannon fire; then come long stretches of minor changes, like the pop of toy guns. There is scaling here, too: If you zoom in on an individual cluster of big changes, you find it is made up of smaller clusters. Zoom again, and you find even finer clusters. It is a fractal structure. Nor is it just the price changes of interest; at times, the price levels also exhibit some kind of irregular regularity. The charts sometimes rise or fall in long waves, or with small waves superimposed on bigger waves. But none of these phenomena—clusters of volatility, or irregular trends—resemble any of the cycles, waves, or other patterns that characterize those aspects of nature controlled through well-established science. There are no familiar sine or cosine waves, with regular periods, of the kind that undulate evenly across the green screen of an old oscilloscope. These peculiar patterns cannot be predicted; and so humans who bet on them often lose. Yet there clearly is a system to them. It is as if the charts have a memory of their past. If the price changes start to cluster, or the prices themselves start to rise, they have a slight tendency to keep doing so for a while—and then, without warning, they stop. They may even flip to the opposite trend.

This is maddening. Our alien, seeing a planet obsessed by so illogical a system, quickly decamps. But his observation of two forms of wildness remain: abrupt change, and almost-trends. These are the two basic facts of a financial market, the facts that any model must accommodate.

Two Dual Forms of Wild Variability

In science, all important ideas need names and stories to fix them in the memory. It occurred to me that the market's first wild trait, abrupt change or discontinuity, is prefigured in the Bible tale of Noah. As *Genesis* relates, in Noah's six hundredth year God ordered the Great Flood to purify a wicked world. Then "were all the fountains of the great deep broken up, and the windows of heaven were opened." Noah survived, of course: He prepared against the coming flood by building a ship strong enough to withstand it. The flood came and went—catastrophic, but transient. Market crashes are like that. The 29.2 percent collapse of October 19, 1987, arrived without warning or convincing reason; and at the time, it seemed like the end of the financial world. Smaller squalls strike more often, with more localized effect. In fact, a hierarchy of turbulence, a pattern that scales up and down with time, governs this bad financial weather. At times, even a great bank or brokerage house can seem like a little boat in a big storm.

The market's second wild trait—almost-cycles—is prefigured in the story of Joseph. Pharaoh dreamed that seven fat cattle were feeding in the meadows, when seven lean kine rose out of the Nile and ate them. Likewise, seven scraggly ears of corn consumed seven plump ears. Joseph, a Hebrew slave, called the dreams prophetic: Seven years of famine would follow seven years of prosperity. He advised Pharaoh to stockpile grain for bad times to come. And when all passed as prophesied, "Joseph opened all the storehouses, and sold unto the Egyptians. . . And all countries came into Egypt to Joseph to buy corn; because that the famine was so sore in all lands." Given the profits he and Pharaoh must have made, one might call Joseph the first international arbitrageur. That pattern, familiar from Hurst's work on the Nile, also appears in markets. A big 3 percent change in IBM's stock one day might precede a 2 percent jump another day, then a 1.5 percent change, then a 3.5 percent move—as if the first big jumps were continuing to echo down the succeeding days' trading. Of course, this is not a regular or predictable pattern. But the appearance of one is strong. Behind it is the influence of long-range dependence in an otherwise random process—or, put another way, a long-term memory through which the past continues to influence the random fluctuations of the present.

I call these two distinct forms of wild behavior the *Noah Effect* and the *Joseph Effect*. They are two aspects of one reality. One, the other, and usually both can be read in many financial charts. They mix together like two primary colors. The red of one blends with the blue of the other, to produce an infinite palette of purples and violets. Evidence so far suggests each market—wheat, cotton, dollar/yen, S&P, or GM—may have a different hue, a different mix of the two forms of wildness.

To measure these two effects, I developed new statistical tools. Some focus on α, the index mentioned earlier. A low-α market would be risky, prone to wild price swings. A market with higher α differs less from the classic coin-tossing market. Other of my statistical tests focus on H, the Hurst coefficient for long-range dependence described earlier. An H of one half implies each price change is independent of the last. A larger H suggests the data are "persistent," trending in the same direction. A smaller H implies "anti-persistence," a tendency to double back on themselves.

To separate the two effects, measured by H and α, I developed a statistical test called rescaled range analysis, or R/S; the name is short for range divided by standard deviation. It is of a type known by statisticians as "non-parametric," tests that make no simplifying assumptions about how the data are organized, and thus do not try to boil everything down to such common parameters as mean and variance that presume a bell-curve distribution. The idea is simple: The Joseph Effect depends on the precise order of events, while the Noah Effect depends on the relative size of each event. Reshuffle the data, like a deck of cards. The cards are all out of sequence now; whatever Joseph Effect was originally present is scrambled out of existence. Only the face value of the cards—the relative size of the events, or Noah Effect—remains visible before and after the shuffling. To complete the test, just compare the deck before and after shuffling. If there is a difference, it must be due to the long-term dependence in the original data; the precise sequence must have been important in the original data, and the degree of that importance can be measured. If there is no before-and-after difference, then whatever dependence was originally present must have been negligible. Result: a measure for long-term dependence.

Now, as fate would have it, under some circumstances these two effects are so closely interrelated that H is simply equal to $1/\alpha$. Take the coin-tossing case: its H is one half and its α is two. Mathematically, the relation between the two effects is quite profound; it presents what mathematicians call a dual relationship.

A Good Reason for "Bubbles"

But how exactly do these two effects—Noah and Joseph, dependence and discontinuity, H and α—interact in markets? Answer: At least one market mechanism I identified naturally leads to the other. Suppose, for instance, that you have an "almost-trend" emerging in a stock price: a few weeks, say, in which a stock price rises seven days out of ten. The pattern must eventually break up, of course; otherwise, it would be a real trend that you could bet on continuing for another few weeks, and hope to make some real money. But when the "almost-trend" finally does break, it can do so rapidly. A sudden lurch downward, perhaps. A discontinuity. Or, in the terms of the Biblical metaphor, a Noah Effect produced by Joseph-style dependence.

For some real-world examples, think about investment bubbles. They can seem calamitous—but they happen all the time, whether in a broad market index like the Dow or in individual assets like a municipal bond. Conventional economics tells us they are aberrations, "irrational" deviations from the norm, caused by a rapacious speculator, mass greed, or some other unpleasant factor. But under certain circumstances they can be entirely rational and flow from the entwined effects of long-term dependence and discontinuity.

Consider the Blowing Bubbles (1) diagram following. Imagine we are following the price of an agricultural commodity, say wheat. Now build a simple, two-part model. The first part calculates the theoretical, "real" value of the coming harvest, per bushel. If the weather is good for a day, the theoretical value should drop slightly. After all good weather presages an abundant harvest and low prices. Say one good day reduces the value by one cent a bushel. In bad weather, the value should rise a cent. In indifferent weather, no change. Now the second part: the actual price in the marketplace. It can easily stray far from the "real" value, as investors place imperfect bets on what the wheat will eventually be worth when harvest finally comes. Of course, by harvest time the two figures, real value and market price, must converge; otherwise, the very real crop will never change hands.

But the price gyrations along the way can be extreme. For instance, if you get a run of bad weather the real value will gradually rise by, say, a cent a day. But the market price will rise even faster; investors anticipate yet more bad weather to come. Then the weather breaks. The price crashes back to the real value, as

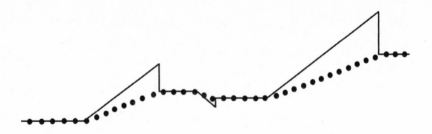

Blowing bubbles (1). It is easy to see how a price bubble develops. The dotted lines trace the theoretical, per-bushel "real" value of a crop as harvest approaches. Each dot represents a day. Every day the weather is bad, the value rises a penny a bushel. The solid line represents the market price. As long as the weather remains bad, the price shoots up as people anticipate more bad weather to come—overshooting the real value. At the slightest break in the weather, however, the price plunges back down as people realize their error.

investors realize they had been over-optimistic and rush to sell before it is too late. That same pattern, of overshooting and crash, recurs—incessantly. Overshoot, then crash, Joseph Effect, then Noah Effect, again and again. How big is the overshooting? It can be estimated after the fact from real supply-and-demand data or in theory from the value of α we find in the market data. But that is little consolation to anyone living in real markets. They cannot forecast the next day's weather with certainty—and so can never forecast when, exactly, the bubble will burst. Result: Prices gyrate, from boom to bust, from bust to boom.

The same saw-tooth price pattern can be imagined for a stock price. Imagine the "crop" is now an industrial company, and the weather is the economic climate that either helps or hinders its growth. Again, prices will overshoot and undershoot. And the longer a company grows, the longer investors will expect it to keep growing.

Did not the recipe behind the Internet bubble somehow reflect this effect? Consider Cisco Systems. The company, the biggest man-

Blowing bubbles (2). Truth is stranger than fiction. This shows how the quarterly stock price of Cisco Systems, the ultimate Internet bubble company, departed from a common measure of its "real" value: its quarterly earnings per share. Just as in the theoretical bubble diagram on the previous page, so in this real chart you can see how enthusiastic investors extrapolated the earnings trends of 1999 into a soaring stock price. In 2000, as earnings flattened, investors started sobering up and the bubble began deflating. And in 2001, when Cisco reported its first quarterly loss, the price fell back to earth.

ufacturer of computers for routing Internet traffic, was viewed as the GM of the Information Age: It made the chassis and engines on which all the rest of the New Economy would drive. It managed an extraordinary record of revenue growth: an average 53 percent annually from 1995 to 2000. And Wall Street came to expect an extraordinary 20 percent profit growth. As investors extrapolated even greater growth, Cisco's stock price soared an average 101 percent a year throughout the 1990s. Its market value hit nearly $500 billion. One bullish brokerage house, Credit Suisse First Boston, issued an investment circular to its clients with the headline: "Cisco—Potentially The First Trillion Dollar Market Cap Company." Of course, the inevitable crash came, and the stock skidded. The chart Blowing Bubbles (2) on page 205 shows the result. Its resemblance to the bubble diagram previous is more than casual. And Cisco investors were not irrational. They saw the company growing, and simply extrapolated that it would continue growing. They knew it would stumble eventually—but when? No way of knowing in a financial market, where long-term dependence and discontinuity combine. The future is shrouded in mist and doubt.

CHAPTER XI

The Multifractal Nature of Trading Time

IMAGINE A CURRENCY TRADER, hunched over a Reuters terminal. The yen and dollar quotes buck up and down, turning green or red. On occasion, trading is fast. Scores of news items are flitting across the electronic "crawl" on the bottom of the screen. Colleagues are waving and shouting all around. Phones are ringing. Customers are zapping electronic orders. The volume of trades is climbing, and prices are flying by. On such days are fortunes won or lost. Time flies.

Then there are the slow times. No news, only tired reports from the in-house financial analysts to chew over. The customers seem to be on holiday. Trading is thin. Prices are quiet. No big money to be made here; might as well go for a long lunch. Time hangs heavy.

Just handy metaphors? Not at all: They are at the heart of how a financial market really works. Imagine for a moment that you could take the tape—the New York Stock Exchange's ticker, or the Reuters record of currency quotes—and play it fast or slow, like a videocassette tape. Run it slowly when prices are flying; there is so much action packed into the tape that you can only see it all by liberal use of the "pause" and "review" buttons. Speed it up during the boring parts, when there is little new information to digest. This is, it turns out, exactly how to analyze a financial market—and exactly how my current and best mathematical simulations of the market work. Their engine is a "multifractal" process: It takes normal clock time, deforms it into a unique form of "trading time," and then generates a price chart from it all. Or it can go in the other direction. It can start with a normal price chart, and break it down into its two primitive components: one process that deforms time and another that generates a price. To what end? To make a lab-bench model of the market that we can use to assess risk, analyze investments, or guard against ruin.

The key to it is multifractals, a subtle and beautiful topic. Recall the definition of a fractal: a pattern or object whose parts echo the whole, only scaled down. By contrast, a multifractal has more than one scaling ratio in the same object—some parts of the object shrink quickly, others slowly. Put it another way. A fractal is like an object defined to be shown in black and white: A point belonging to the fractal set is shown in black and a point that does not belong is left white. A multifractal takes this to the next level, to objects that involve halftones, shades of gray. Since the world is not black and white, the study of multifractals comes closer to the way many aspects of nature really work. It is the way gold ore clusters here and there on the surface of the earth; the way oil reserves appear to concentrate in certain strife-prone parts of the world; the way the velocity of the wind on a stormy day comes "intermittently," in clusters of high gusts, interspersed with gentler breezes.

And it is also the way price-changes in a financial market can cluster into zones of high drama and slow evolution. I began my research in finance with cotton and the Noah Effect—the wild price swings and "fat tails." I continued a few years later, as a byproduct of the Nile floods, with the Joseph Effect—the interdependence of price changes across time, or "long memory." The next advance came, again a few years later, from the study of wind: the intermittence of turbulence. To describe the path I followed to multifractality as the tool for this study would take us too far afield but it had familiar ingredients. The simulation of wind-gust, first published in 1972 and reproduced in Chapter 6, reminded me of how the volatility of cotton prices varies over successive months. A second key was analytic: My earliest multifractals have tails that follow a power-law distribution. Shortly afterward I found that all multifractals manifest a Noah Effect that can vary over a broad range of differing degrees: Change can be sharp and violent—whether a burst of sunspot activity, or a crash on the New York Stock Exchange. In addition, every multifractal manifests the Joseph Effect: Every part of the object under study, whether a map of galaxy clusters or a record of T-bill rates, influences every other part. Brought together in a multifractal model, the Noah and Joseph Effects hold a mirror up to the market, revealing it to be both highly risky and subtly interdependent.

As a theory of the real world improves, it moves on from black-and-white to shades of grey. Therefore, as early as 1975, I extended to all fields the notion that to improve almost any fractal model it is a good idea to replace it by a multifractal one.

Looney 'Toons for the Last Time

Keep it simple is the catchphrase of good models. So we come back to the fractal cartoons, our sketches of how the basic concepts of fractal market analysis come together. The goal is to simplify but not oversimplify. Therefore the cartoons are meant to be less realistic than my preferred model, yet able to be tuned to capture the essence of every type of market effect, from Bachelier's original idea of a Brownian motion, and then to Noah, Joseph, and both together.

First, let us pull together all the strands of the prior cartoons—a recapitulation of old themes. As shown earlier, the financial cartoons begin with a simple seed and build to complexity. We start with a rising, straight line in a box. Next comes the generator, a

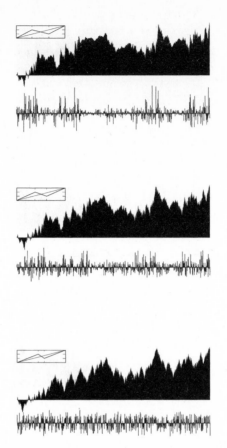

Panorama of financial multifractals. As shown several times, fractal cartoons can simulate price charts of different stripes. These two pages recapitulate the by-now familiar Brownian model (bottom left) and introduce five variants. Each variant is obtained by changing the shape of the zigzag generator (shown in the inset boxes.) By sliding the two break points horizontally, so that they come closer together or spread farther apart, the resulting fractal fever chart changes—and

lightning-bolt shape, that fits over the straight line. Then, wherever a straight line appears, interpolate a small copy of the generator in its place. Repeat, at ever-smaller scales. Gradually, a mechanical sort of zigzag chart takes shape.

As shown before, the magic begins when you play with the generator. Its precise shape matters greatly to the outcome. You can

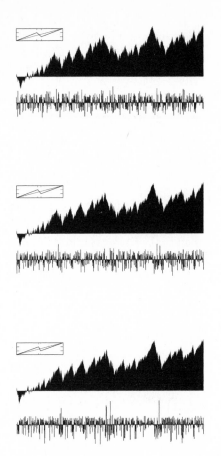

each price-difference diagram reflects the changes dramatically. When the break points are farthest apart, as in the top left panel, one gets extremely wild variation—just like a real price chart. When they are closest together, as in the bottom right panel, the variation is still wild—but less realistic. The theory of multifractals has tools to characterize the differences among the diagrams.

change the number of points where it zigs and zags, from two, to one, three, or any other number. You can change the coordinates of the break-points. The Brownian chart kept a precise relationship between the width and height of each interval in the generator: It was, you recall, linked by a power law in which one was the square root of the other. But by choosing a different power, we could generate charts that showed varying degrees of long-term dependence.

And by adding vertical jumps, we could produce charts that showed the fat tails and discontinuity of cotton prices.

In fact, you can make an infinite variety of charts—and I urge you to try it yourself, with the computer or even (but it takes more time!) with paper and pencil. Pick a shape for a generator, interpolate it, and see what kind of diagram emerges. Many will be extravagantly messy, far messier than anything you could recognize as a "real" price chart. But many others will be quite realistic. In fact, the variety of the results is such that I had at an early stage to organize them and show how the different generators and diagrams relate to one another. The flexibility of the method can be seen in the two-page diagram —a "Panorama" of one family of financial multifractal cartoons. Focus on the generators shown, sow in the inset boxes. As you change the distance between the break-points, the generator will change systematically—as will the shape of the fractal price charts that result.

Multifractal Time

How do these generators relate to one another? In some cases, quite intimately. It fact, you can design a generator that, in a sense, inherits characteristics of two other generators. In the diagram following, I show how two parent generators—a "father" and a "mother"— come together to produce a new "baby" generator that partakes somewhat of the traits of both. A mathematical game? Not at all. As will be seen, it lends to new versatility in producing financial fractals.

So what is going on here? Look at the axis labels on the diagram: t, θ, and P. The first stands for clock time; the last is for price; and the middle one, labeled as the Greek letter theta, denotes an auxiliary scale called "trading time." In summary: The family starts with the parents. The father takes clock time and transforms it into trading time. The mother takes clock time and changes it into a price.

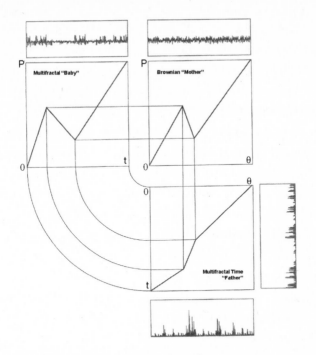

The Baby Theorem. This diagram shows how two generators can pass on traits to a third. The mother generator at top right is a Brownian motion, in conventional clock time—as apparent from the chart of its increments shown above the generator. The father, at bottom right, transforms clock time into a new time-scale, called trading time. By adopting the father's trading time, the mother creates a multifractal baby (top left). Baby's increments, shown above its generator, would pass the "find the fakes" test with flying colors: It is, to all appearances, a genuine price chart. Meanwhile, the uneven, slow-and-fast nature of trading time is shown in the two time-increment charts to the father's bottom and right. And as in the previous, two-page illustration, the horizontal displacement of the generators' break points is the critical step in this particular fractal process. Broadening the gap between the mother's break points yields the baby's generator. I called it the Baby Theorem at first because its mathematical proof was easy, even if its consequences are far-reaching…a common occurrence in science.

Merged together, the baby takes the father's trading-time and converts it into a price by the rules the mother provides. Last step: Use the new, baby generator to make a full fractal price chart that is a variant of one of the panels in the "Panorama of financial multifractal." And there you are: a realistic financial chart, made by stretching and shrinking time. And a nice metaphor for our age, some fifty years after the discovery of the double helix: Each parent contributes one half of a chromosome to the baby.

You can see how all this fits together, on paper, in the "fractal market cube" diagram on page 214, a three-dimensional sketch of the price-generating process. On the left sidewall is the fractal chart that the mother generator produces. It is a variant cartoon of the Brownian motion model of how prices happen—in fact, a cartoon pared to the essentials of our original, up-down-up generator without any random shuffling. This explains its well-behaved appearance. The fractal chart produced by the father fluctuates sedately

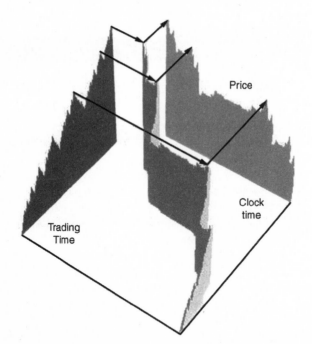

The fractal market cube. Two processes come together to produce a financial chart. This 3D cube shows how. It is, in fact, just a different way of representing what was shown in the prior, Baby Theorem diagram. The left wall is a non-randomized cartoon of Brownian motion—a variant of the mother fractal. The jagged path along the floor is the father; it shows clock time getting deformed, in fits and starts, into a new scale of time, trading time. On the right wall is the baby, the merged chart of multifractal price versus clock time.

along the floor, zigzagging around the diagonal. He converts clock time into a multifractal trading time in an erratic process; his videotape of time is speeding up and slowing down, in fits and starts. Lines drawn sideways from the mother and vertically from the father meet at the top—and then project, as the arrows show, along the right-side wall—the baby price chart. This is the final financial chart. It fluctuates wildly. It has the big jumps and "fat tails" we find in real price charts, as well as the long-term dependence and persistence of the real thing. The baby looks, for all the world, like a stock price or exchange rate. To switch metaphors: It is

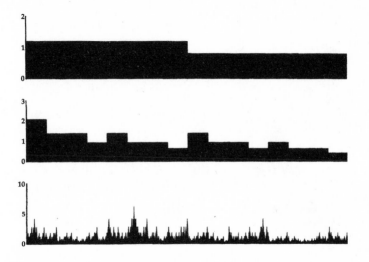

The binomial bending of time. How do you deform time? The father, in the previous diagrams, is a mathematical process called a multiplicative cascade. A simple example of such a cascade is shown in this diagram. Consider, at successively finer scales, a cross-section cut through a gold-producing country. The top rectangle is the first approximation. It shows 60 percent of the gold is to the left and 40 percent to the right. Then look in finer detail. Cut each half of the map in two halves, with 60 percent to the left and 40 percent to the right. Keep doing this. The outcome is in the final panel: The original area is partitioned irregularly, with tall peaks and low valleys—places of high and low ore concentration. Now think of the substance being divided as time, not gold. Time would bunch and move quickly at the peaks, thin and slow in the valleys. This is the essence of the time-deformation process shown in earlier cartoons as the father.

not a mixture, but an alloy of its two parent metals, like brass smelted from tin and copper. And like an alloy, its behavior is different from that of either precursor metal alone.

But how, you may wonder, does the "father" chart actually deform clock time? Its mathematical engine is called a multiplicative cascade—a fancy term for a type of fractal process entailing many repeated multiplications. Imagine time as if it were a form of matter: This being a book on finance, think of it as if it were gold ore. Remember: fractals are not about the "things" themselves but about their common property of roughness. This is not a farfetched but an apt idea, because, obviously, gold is not distributed evenly around the world. It clusters here and there—just the way the action in a financial market clusters into different stretches of time. We can mimic that effect mathematically. Pull out a map of gold-rich South Africa, specifically, a cross-section of the earth there along a west-to-east line. Start with a low-resolution map that divides the country into two pieces, one east and one west. About 60 percent of the gold ore lies in the western half, and 40 percent in the east. Look more closely: Cut each half into halves again. Finer processes concentrated 60 percent of the western gold into the westernmost quarter—or 36 percent of the total gold deposits (60 percent times 60 percent equals 36 percent of the total). Forty percent of the western gold is in the second quarter of the map; that means 24 percent (40percent times 60percent equals 24percent) of all the ore lies there. Continue on, multiplying again and again, re-partitioning the cut, redistributing the gold across the entire interval. The result is plotted on page 215: it is a very uneven distribution: Some parts of the cut are rich in ore, others not worth a prospector's visit.

The same kind of mathematics can be turned to bunching time into irregularly spaced segments. In fact, this concept of trading time predates multifractals, originating in a paper I coauthored in 1967. It remains mostly speculation. But it already permits some extraordinarily faithful reproductions of a financial market.

Beyond Cartoons: The Multifractal Model with No Grids

To repeat: Both for the prices and trading time, the cartoons' virtue is that they replace reality by something simple and easy to manage. But there is no free lunch and simplification has a cost. Instead, it is best to go beyond cartoons. My current best model of how a market works is fractional Brownian motion of multifractal time. It has been called the Multifractal Model of Asset Returns. The basic ideas are similar to the cartoon versions above—though far more intricate, mathematically. The cartoon of Brownian motion gets replaced by an equation that a computer can calculate. The trading-time process is expressed by another mathematical function, called $f(\alpha)$, that can be tuned to fit a wide range of market behavior. My model redistributes time. It compresses it in some places, stretches it out in others. The result appears very wild, very random. The two functions, of time and Brownian motion, work together in what mathematicians call a compound manner: Price is a function of trading time, which in turn is a function of clock time. Again, the two steps in the model combine to produce a "baby" far different from either parent.

The final product has wild price fluctuations—the big jumps and "fat tails" we saw in cotton and many other non-normal price charts. It has the volatility that clusters here and there: Periods of big price changes group together, interspersed by intervals of more sedate variation—the tell-tale marks of long memory and persistence. It shows scaling: The "moments," a term for the basic statistical characteristics of the price series, follow a familiar scaling pattern that is now captured in the function $f(\alpha)$. In fact, you can fashion an entire spectrum of price charts, some very wild, some very dependent, some partaking of both—just as we did earlier with the simplistic cartoons.

Research so far suggests the model is accurate. In the late 1990s, the first tests of the model's applicability were conducted in the doctoral dissertations of two of my Yale students, Laurent Calvet and Adlai Fisher, now teaching at, respectively, Harvard University and the University of British Columbia. We focused on the global market for dollar-Deutschemark exchange. It, like cotton, has special appeal to an economist. Its volume is huge. Its significance to the global economy is great. And its records are long, copious, reliable, and easily available. We used the real data of the real marketplace: tick-by-tick records, from thousands of trading screens across the world, of the live quotes posted by banks and other major currency traders. These data, gathered and stored by a Zurich consulting firm, Olsen & Associates, focused on a one-year period from autumn 1992 to autumn 1993—1,472,241 prices in all. For easier comparison with other economic research, we also looked at a conventional data set: twenty-four years of daily dollar-Deutschemark quotes at 4 p.m. London time, from 1973 to 1996.

The model passed the test. Price changes in this currency market clearly do scale as the model predicts. Volatility clusters. Episodes of fast action intersperse with intervals of slow, dull trading. Zoom in on the fast episodes, and they are seen to have sub-clusters of fast and slow sub-intervals—clusters within clusters within clusters. It is a classic multifractal pattern.

Its scaling stretches, through every focal length of our mathematical zoom lens, from about two hours to 180 days—an unusually long zone of regularity. At shorter time-intervals, a new pattern emerges: What economists call market "microstructure" starts to kick in. Here, the average price change is up or down by just 0.14 pfennig, only twice the spread of 0.7 pfennig between bid and ask. With such narrow profit opportunities, some traders do not bother changing their quotes instantly, so you would expect the data to look differently. At intervals longer than 180 days, yet another effect alters the data stream. The Noah Effect is fading. The wild price variability is settling down. These two bounds, below two hours and above 180 days, are called crossovers: points where a new mathematical relation takes hold. Crossovers are common in real, as opposed to theoretical, fractal data. Consider a typical real fractal: the way air passages in the lungs branch from the main bronchial tubes to the millions of tiny bronchii feeding individual alveoli cells. There is a physical limit to how many and how tiny these fractal tubes can be, or need to be, for the support of life. Tubes above and below a certain size, the crossovers, simply do not occur in nature. Likewise with financial data. Scaling works in the broad, macroscopic middle of the spectrum; but at the far ends, in what you might call the quantum and cosmic zones, new laws of economic life apply.

As always, one set of data is—well, just one set of data. Could our tidy results be a mere fluke of the specific currency records we analyzed? No. To guard against that, we also ran our equations over another set of dollar-Deutschemark prices, this time from the U.S. Federal Reserve Board. Same results: It works. What if the model only works for that one market? No again. We began testing other markets. The degree of "fit," as economists call a result that matches a model's expectations, varied—as it often does in statistical research. Some assets fit our scaling model perfectly. Stock in Archer Daniel Midlands, Lockheed, Motorola, and UAL were textbook multifractals. Stock in General Motors, a broad index of U.S. stock prices, and the dollar-yen exchange rate were also multifractal—though over a narrower range of time-scales.

At least as important: The model successfully solved several old problems that had bedeviled my prior research. From the very beginning, in 1963, some economists had pointed out that the degree of wildness—the fatness of the tails—appeared to diminish as you looked at returns over longer and longer time-periods, from a day to a year to a decade. The common wisdom in economics was, and in some circles still is, that I may be right that daily or weekly prices do not follow the standard model, but who cares? Most people, goes the argument, buy and hold for months, years, or decades—and in those time-scales, the conventional models work just fine. There is a fallacy in this, of course. Most people also do not contract HIV and then develop AIDS, but the few percent who do get it are very glad that the pharmaceutical industry has taken the time and expense to develop the necessary drugs to keep them alive longer. More importantly, the multifractal model successfully predicts what the data show: that at short time-frames prices vary wildly, and at longer time-frames they start to settle down.

Putting the Model to Work

But enough of theory. How do you use these ideas as a real-world financial tool? First, the equations need to feed into a computer model. The model must work two ways, forward and backward. Forward means that we should be able to construct artificial price charts from the fractal seeds, just as we did with the cartoons. Backward means that we should be able to take raw price data, analyze it on our computers, and estimate the key parameters that the multifractal model requires. Then using those values, we should be able to tell the computer to reconstitute the market—to generate an artificial price series that differs from the real one but follows the same statistical pattern.

That is exactly what we have done, repeatedly, using a common computer technique called a Monte Carlo simulation. The result was excellent forgeries of the market—not identical, but statistically similar to the genuine article. What good is a forgery, you may ask? An explanation is in order. Whenever you compress data—whether a computer file or a price series—you reduce it to fewer pieces of information, to a small number of parameters. Then when you decompress it again, you do not get the full set of data back again; instead, you get something that is close enough to the original for whatever purpose you have. For instance, a Cartier-Bresson photograph can be compressed for e-mailing to someone, then reconstituted upon receipt into something that is grainier than the original photograph—but not

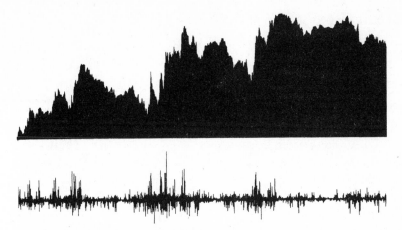

And here's one I made earlier.... This is an example of the final product of my Multifractal Model of Asset Returns: A very faithful copy of a real price chart (with the chart of price changes below it). If it looks familiar, it should: It was the model used in the Spot-the-Fakes contest in the opening chapter.

noticeably so on a normal computer screen. It is "good enough" for the purpose at hand. In the same way, in financial modeling all we need is a model "good enough" to make financial decisions. If you can distill the essence of GE's stock behavior over the past twenty years, then you can apply it to financial engineering. You can estimate the risk of holding the stock over the next twenty years. You can estimate how many shares of the stock to buy for your portfolio. You can calculate the proper value of options you want to trade on the stock.

This is, of course, exactly the aim of all financial theory, conventional or not. The one difference: This time around, it would be nice to have an accurate model.

To me, the greatest charm of the multifractal model is its economy. One simple set of rules can produce a great variety of behavior, depending on the circumstances. By contrast, most financial academics are going through a love affair with another way of modeling market volatility. Its main inventor, Robert F. Engle, shared a Nobel in 2003 for its development. It starts from some of the same facts I have been advancing in this book: Volatility clusters, due to dependence. To model that, it has already been mentioned that a set of statistical tools was developed; it is called GARCH, short for Generalized Auto-Regressive Conditional Heteroskedasticity. To model the clusters, it starts with a conventional Brownian model of price variation. When the volatility jumps, it plugs in new parameters to make the bell curve grow; when the volatility falls, new parameters shrink the curve. You might say the bell vibrates, to fit the circumstances. GARCH is, certainly, a handy abacus now used by many options traders and finance directors trying to model risk. But it begs the question of what makes the bell vibrate. And, as you try to work with the model, it becomes increasingly complicated.

To say much with little: Such is the goal of good science. But most established financial models say little with much. They input endless data, require many parameters, take long calculation. When they fail, by losing money, they are seldom thrown away as a bad start. Rather, they are "fixed." They are amended, qualified, particularized, expanded, and complicated. Bit by bit, from a bad seed a big but sickly tree is built, with glue, nails, screws, and scaffolding. That people still lose money on these models should come as no great surprise.

The multifractal model, by contrast, begins with the unchanging, fundamental facts of market behavior—the "invariances," a mathematician would call them. It is economical and flexible and mimics the real thing. In designing models, I think back to the great exemplars of history. Consider Newton's famous law of gravity: The force of attraction between two bodies depends on their distance. He needed just a few pen strokes to express that thought, mathematically. But from it, he showed why the planets move as they do, where comets fly—even how high the tides flow. Later generations elaborated, until we had rockets, satellites, and men in space. His was a very small seed of thought, from which a great forest of science and engineering has grown. My hope is that, some day, the small seed of multifractal analysis can grow into a fruitful new way of managing the world's money and economy.

PART THREE

· · · ·

The Way Ahead

Pharaoh's breastplate. (Overleaf.) Cover of Mandelbrot 1999a. Illustration of a fractal structure made of an infinity of circles. It is called the limit set of a Kleinian group—another example of the power of very simple fractal formulae to create ordered complexity.

CHAPTER XII

Ten Heresies of Finance

FOR A REAL GRASP of economics, skip the books and lectures. Get into the garment trade.

In 1945, my father tried to restart the clothing business he had before the war. These were, of course, hard times. The rubble of war lay all around: factories broken, commerce disrupted, lives shattered, food rationed. And warm clothing was scarce. So my father traveled from the great city of Paris to the sheepfolds of the Massif Central, and there bought cheap, rough wool cloth from the small mills. He brought it back to our house in the down-at-heels Nineteenth Arrondissement where he cut it into patterns. I, with my long arms and young steady hands, helped him whenever I visited from school. Then a young man with a truck came from a far suburb to collect the pieces for an aunt, a mother, or a concierge to stitch together. And at last the *culottes* or *blousons* returned, completed, to my father for re-sale.

But at what price? That depended not on my father's cost, but on whatever value people saw in the garments, and that value blew in the wind. At the start, the business did well enough. But my father died; tastes changed; and coarse, hand-sewn woolens were no longer the blessing they had once seemed. Suddenly, my father's inventory had little value. Several merchants came to buy the stock, but my mother refused all the offers; either out of respect for my father's memory or out of her own stubbornness, she would not sell below cost. Finally, I took matters into my own hands and one day while my mother was away sold it off. From it, I got a cleared-out room for other uses, some extra cash for the family—and a lively appreciation for the slipperiness of that classic economic concept, value.

Much of what passes for orthodoxy in economics and finance proves, on closer examination, to be shaky business. Since my youth, I have been shamelessly disrespectful of received wisdom. I question those who tell me such a thing is possible or such another is impossible. How would they know? Have they tried it themselves? My understanding of economics comes not from abstract theory, but from observation. Though I later lectured on economics at Harvard, I did not begin its serious study until I was nearly thirty years old, well after my training as a mathematician and scientist has left its mark. Rather, my approach has been that of a practical man, a practicing scientist, an objective observer of what actually happens in a financial market rather than of what people believe or wish to happen.

What I have found in finance is a collection of—to me—obvious facts. Some have fed into my fractal analysis of the market; others are deduced from it. That they often contradict received wisdom, I cannot help. But, given how finance is organized today, it can only give me hope to find myself so often proceeding contrary to dogma. Though discussed piecemeal, here and there in earlier chapters, a summary may be helpful now. Hence, in the "list" style of a newspaper columnist, I present my Ten Heresies of Finance.

. . .

1. Markets Are Turbulent.

To truly understand something, you must experience it—get it under your fingertips. When I started studying turbulence forty years ago, still at Harvard and deeply involved with cotton prices and the Nile, I also started studying turbulence. The trigger was a lecture by Robert W. Stewart, a professor at Vancouver with a trove of data on the subject. Researchers had fitted an old, surplus submarine with a long snout, and fixed recording apparatus at its tip. Then they had piloted the sub slowly through the wild crosscurrents, eddies, and vortices of Puget Sound. The result: a rich harvest of data on turbulence in water.

On a visit to Vancouver, I asked to listen to the recordings. Not possible, I was told; the audio tapes, while playable, spanned too broad a frequency spectrum from high pitch to low, most of them outside human earshot. But surely, I said, you can speed up and slow down the tape? I insisted. And, after some fumbling with the then-primitive equipment, they obliged me. We sat and listened. Just listened. Loud high pitch, then low rumblings. Then high pitch again; more rumblings. Change the tape speed: Same pattern. Now, most people listening to this would call it stretches of high-frequency noise interrupted by low patches. But if they had taken the trouble to study the intervals, to analyze the relative proportions of high and low patches, they would have found something else: a turbulent process that proceeds in bursts and pauses, and whose parts scale fractally. The turbulent water through which the submarine's nose plowed in a one-dimensional line was not one long alternation of fast and slow water. Instead, seen in all three dimensions, it was a complicated pattern of churning eddies and torrents, all interrelated from start of journey to end of journey—in effect, over an infinite span of time and space.

That experience underlies all my thinking about financial markets. The tell-tale traces of turbulence are plainly there, in the price charts. It has the turbulent parts that scale up to echo the whole. It has a set of numbers—a multifractal spectrum—that characterizes the scaling. It has a long-term dependence so that an event here and now affects every other event elsewhere and in the distant future. It shows turbulence in a wild kind of variation far outside the normal expectations of the bell curve; in a concentration of changes here and there; in a discontinuity in the system jumping from one value to another; and in one set of mathematical rules that can, in large measure, describe it all. This is a lot to assert, and as this book has proceeded, the evidence and theory have appeared bit by bit. But it all comes together in the metaphor of turbulence.

Why are markets turbulent? I am a scientist, not a philosopher; so I can only hazard some suggestions. One possible source is the world outside the markets—what economists call exogenous effects. After I had, in the early 1960s, focused on scaling and long-term dependence, key traits of turbulence, I soon found innumerable other examples in many natural and economic phenomena; these phenomena, in turn, may impress a corresponding pattern on prices. For instance, I have found characteristic scaling patterns, from many small items to a few large ones, in the area and reserves of oil fields. The valuation of certain gold, uranium, and diamond mines in South Africa scales. Storms and earthquakes scale.

You can imagine a chain reaction. Weather affects harvests, and harvests affect prices. The distribution of natural resources around the globe—oil, gold, and other minerals—affects supply, hence affects prices. The same goes for business: The size of firms in an industry, from a mighty Microsoft to a legion of little software houses, also follows a scaling pattern. So, industry concentration affects profit, hence affects stock prices. Now, this is unsatisfactory for a rigorous analysis of cause and effect in economics. But if one must have a "story" to explain the data, then this is at least a plausible partial one. Scaling enters the system from the fundamentals of weather patterns, resource distributions, and industrial organiza-

tion. Scaling finishes—and feeds back through the system again—
in the marketplace.

Similarly, long-term dependence, another characteristic of turbulence, is found all round us. Think of a small country, like Sweden,
where every big company does business, directly or indirectly, with
every other one. Volvo does something that affects Saab—say,
launches a new car model that steals market share. Saab comes back
with a fancier car, making satellite-location services standard rather
than an expensive option, and so Ericsson starts selling more Global
Positioning System receivers. And so it spins on, throughout the
Swedish economy—and spilling gradually into neighboring
Finland and Nokia, to Norway and Statoil, and as far around the
globe in ever-diminishing ripples as we can measure it. Now imagine the same phenomenon in a large country, like the United States.
How much more numerous, more complex, more significant are
the economic repercussions of any one company's actions? Imagine,
finally, the world economy: a chamber of mirrors. Each company
relays, distorts, and attenuates the economic signals as they flash
around the globe. The signals fade in time. But it can take months,
years, or decades for a signal to become so weak and remote as to be
unremarkable. Such is long-term dependence in an economy: Every
event, no matter how remote or long ago, echoes across all other
events.

No question, such speculation is very tentative, and I prefer to
avoid it. To drive a car, you do not need to know how it goes; similarly, to invest in markets, you do not need to know why they
behave the way they do. Compared to other disciplines, economics
tends to let its theory gallop well ahead of its evidence. I prefer to
keep theory under control and stick to the data I have and the mathematical tools I have devised. They permit me to describe the market in objective and mathematical terms as turbulent. Until the
study of finance advances, for the how and why we will each have to
look to our own imaginations.

. . .

2.   Markets Are Very, Very Risky— More Risky Than the Standard Theories Imagine.

Turbulence is dangerous. Its output—the pressure or velocity of water, the average or change in price—can swing wildly, suddenly. It is hard to predict, harder to protect against, hardest of all to engineer and profit from. Conventional finance ignores this, of course. It assumes the financial system is a linear, continuous, rational machine. That kind of thinking ties conventional economists into logical knots.

Consider the so-called Equity Premium Puzzle, a chestnut of the scholarly literature since its discovery two decades ago by two young economists, Rajnish Mehra and Edward C. Prescott. Why is it that stocks, according to the averages, generally reward investors so richly? The data say that, over the long stretch of the twentieth century, stocks provided a massive "premium" return over that of supposedly safer investments, such as U.S. Treasury Bills. Inflation-adjusted estimates of that premium vary, depending on the dates you examine, between 4.1 percent and 8.4 percent. Conventional theory calls this impossible. Only two things, the theory says, could so inflate stock prices: Either the market is so risky that people will not invest otherwise, or people merely *fear* it is too risky and so will not invest otherwise. Now, when studying this, economists typically measure the real market risk by its volatility— quantified by their old friend, the bell-curve standard deviation. They measure people's perception of risk from opinion surveys. Then they do the math, and come up short: The conventional formulae say the risk premium should not exceed 1 percent or so. Surely some mistake in the data?

Such was the view of the economics establishment, when Mehra and Prescott first raised the issue. It took them seven years, until

1985, to get their paper past the gatekeepers of the scholarly economics journals. Since then, scores of papers have been written trying to explain the problem away. But these papers miss the point. They assume that the "average" stock-market profit means something to a real person; in fact, it is the extremes of profit or loss that matter most. Just one out-of-the-average year of losing more than a third of capital—as happened with many stocks in 2002—would justifiably scare even the boldest investors away for a long while. The problem also assumes wrongly that the bell curve is a realistic yardstick for measuring the risk. As I have said often, real prices gyrate much more wildly than the Gaussian standards assume. In this light, there is no puzzle to the equity premium. Real investors know better than the economists. They instinctively realize that the market is very, very risky, riskier than the standard models say. So, to compensate them for taking that risk, they naturally demand and often get a higher return.

The same reasoning—that people instinctively understand the market is very risky—helps explain why so much of the world's wealth remains in safe cash, rather than in anything riskier. The Wall Street mantra is asset allocation: Deciding how to divide your portfolio among cash, bonds, stocks, and other asset classes is far more important than the specific stocks or bonds you pick. A typical broker's recommendation, based on Markowitz-Sharpe portfolio theory, is 25 percent cash, 30 percent bonds, and 45 percent stocks. But, according to a study by the Organization for Economic Cooperation and Development, most people do not think that way. Japanese households keep 53 percent of their financial assets in cash, and barely 8 percent in shares (the balance is in other asset classes). Europeans keep 28 percent in cash, 13 percent in shares. For Americans, it is 13 percent cash and 33 percent stocks. Unlike a broker, most investors do not care about "average" returns. For them, the rare, out-of-the-average catastrophes loom larger. Common sense and folk wisdom are often wrong, of course, but must never be ignored.

The ultimate fear is financial ruin. Now, ruin is a much-studied term in risk analysis. It occurs when some measure of wealth—the size of your stock portfolio, the capital reserve account of a bank, the profit or loss of an insurer—falls below some desired threshold. You can calculate the odds of that happening. In the charts following, from a book by Paul Embrechts of the Swiss Federal Institute of Technology and colleagues, you can see several computer simula-

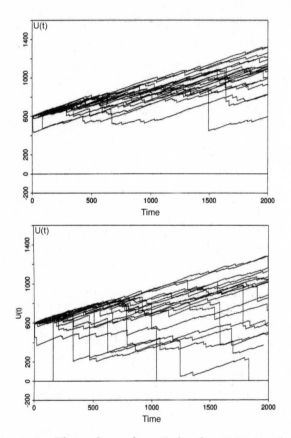

Roads to ruin. These charts, from Embrechts 1997, simulate the profits or losses of several different insurers collecting premiums and paying claims; each line from left to right through time shows the outcome for a different hypothetical insurer. The top chart shows what happens if you assume claims follow a nice, mild pattern of variation: a hugely profitable industry. The bottom chart shows what happens if you assume wild variation in claim sizes: some bankruptcies. Which is closer to reality?

tions of insurer profits. The steady collection of premium income, in the absence of any claims, would keep each simulated company's profits on a nice, steady, rising path. By contrast, when a claim is filed, each company sees its profit lurch downward. If it hits zero, the insurer is ruined. That is the basic premise; the two charts show what happens if you change your assumptions on how the insurance market works. The top diagram: The size and frequency of claims follow a bell-curve. The bottom diagram: A more realistic, scaling probability for claims. The first would have nearly every insurer prospering; the second shows some real-world bankruptcies. A risky business, insurance.

The same kind of simulations can be done for stock, bond, or other financial prices. According to the standard model of finance, in which prices vary according to the bell curve, the odds of ruin are about 10^{-20}. Translation: One chance in a hundred billion billion. With odds like that, you are more likely to get vaporized by a meteorite landing on your house than you are to go bankrupt in a financial market. But if prices vary wildly, as I showed in the cotton market, the odds of ruin soar: They are on the order of one in ten or one in thirty. Considering the disastrous fortunes of many cotton farmers, which estimate of ruin seems most reasonable?

3. Market "Timing" Matters Greatly. Big Gains and Losses Concentrate into Small Packages of Time.

Concentration is common. Look at a map of gold deposits around the world: You see clusters of gold veins—in South Africa and Zimbabwe, in the far reaches of Siberia and elsewhere. This is not total chance; millennia of real tectonic forces gradually worked it that way. Understanding concentration is crucial to many businesses, especially insurance. A recent study of tornado damage in

Texas, Louisiana, and Mississippi found 90 percent of the claims came from just 5 percent of the insured land area.

In a financial market, volatility is concentrated, too; and it is no mystery why. News events—corporate earnings releases, inflation reports, central bank pronouncements—help drive prices. Orthodox economists often model them as a long series of random events spread out over time. While they can be of varying importance and size, their assumed distribution follows the bell curve so that no single one is preeminent. What sense is this? The terrorist attack on the World Trade Center was, by anyone's reckoning, far and away the most important event in years for world stability and, consequently, for financial markets. It forced the closure of the New York Stock Exchange for an unprecedented five days, and when trading reopened caused a 7.5 percent fall. It was one titanic event, not the sum of many small ones. Big news causes big market action. And that action concentrates in small slices of time.

The data demonstrate this. From 1986 to 2003, the dollar traced a long, bumpy descent against the Japanese yen. But nearly half that decline occurred on just ten out of those 4,695 trading days. Put another way, 46 percent of the damage to dollar investors happened on 0.21 percent of the days. Similar statistics apply in other markets. In the 1980s, fully 40 percent of the positive returns from the Standard & Poor's 500 index came during ten days—about 0.5 percent of the time.

What is an investor to do? Brokers often advise their clients to buy and hold. Focus on the average annual increases in stock prices, they say. Do not try to "time the market," seeking the golden moment to buy or sell. But this is wishful thinking. What matters is the particular, not the average. Some of the most successful investors are those who did, in fact, get the timing right. In the space of just two turbulent weeks in 1992, George Soros famously profited about $2 billion by betting against the British pound. Now, very few of us are in that league, but we can in our modest way take cognizance of concentration. Suppose big news has inflated a stock price by 40 percent in a week, more than twice its normal volatility. What are the odds that, anytime soon, yet another 40 percent run will occur? Not impossible, of course, but certainly not large. A prudent investor would do as the Wall Street pros: Take a profit.

4. Prices Often Leap, Not Glide. That Adds to the Risk.

A favorite pastime of cranks and academics is devising the financial equivalent of a perpetual motion machine.

One day when I was working in IBM's lab, I got an urgent order from the top. The company president, Albert L. Williams, had heard at a cocktail party that some MIT professor had found a systematic way to beat the stock market. Williams told somebody, who told somebody who told somebody who told me: Check it out. So I did. The industrial management professor, Stanley S. Alexander, had in 1961 published a scholarly article on a seemingly sure-fire way to get rich quick. He called it a "filter method." In brief: Every time the market rises by 5 percent or more, buy and hold. When it falls back 5 percent, go short and hang on. The point, Alexander argued, is that the orthodox "efficient market" theorists are wrong and prices do tend to move in trends; if a stock rises 5 percent, it is more likely to keep rising than it is to fall. So a simple rule like his could profit from this tendency. And profit hugely: He calculated that an investor who had blindly followed such a rule from 1929 to 1959 would have gained an average 36.8 percent a year, before commission. That was twelve times the average 3 percent increase that the market actually achieved during that period. He concluded, a bit smugly: "I leave to the speculation of others the question of what would happen to the effectiveness of the filter technique if everybody believed in it and operated accordingly."

Well, I pondered it. I banged out a letter to the great professor, on my portable typewriter. I was too low on the IBM totem pole to have anybody type for me. Which of several possible prices, specifically, had he used in his calculations, I asked? He answered, with a dismissive, hand-written scrawl at the bottom of my letter: "It doesn't make any difference."

It certainly did. It made the difference between a 36.8 percent profit and a loss of as much as 90 percent of the investor's capital. The problem was simple. Alexander had calculated the value of his theoretical portfolios using the published daily closing prices, rather than a real-time ticker-tape such as a live investor would encounter. If GM stock rose 6 percent from one day's close to the next, Alexander assumed the investor would have bought on the way up, at precisely the 5 percent mark the filter rule required. In fact, prices do not rise smoothly from one cent to the next; they can easily jump many notches at a time. The precise target, 5 percent, would get bypassed on the rapid rise. The real purchase might not get executed until prices had already climbed 5.5 percent— thereby costing the investor half his potential 1 percent gain on that particular trade. The same thing happens as prices fall: Rather than sell at precisely a 5 percent drop, the investor might actually execute the sale at 5.5 percent—costing him the other half-percent profit Alexander had assured him. The real world clipped his profits on the way up, and stretched his losses on the way down.

In short, like all perpetual motion machines, this one was fatally flawed. I sent a memo back up the IBM command chain to Williams, and never heard back; he may have tried it himself and given it up as a bad bet. But three years later Alexander retracted. In another scholarly article, he reported that most of the profits in his prior portfolios had vanished—and in many cases had swung to a loss—when a more realistic price series was used. "The big, bold profits of Paper 1 must be replaced with rather puny ones," he wrote. "I must admit that the fun has gone out of it somehow."

Alexander can be forgiven the mistake. Continuity is a common human assumption. If we see a man running at one moment here and a half-hour later there, we assume he has run a line covering all the ground in between. It does not occur to us that he may have stopped to rest and then hitched a ride. The greatest innovation of seventeenth-century mathematics, the calculus, was designed to study continuous change; its co-founder, Gottfried von Leibniz, believed deeply in what he called a "principle of continuity." Economists often do the same. Continuity is a fundamental assumption of conventional finance. The mathematics of Bachelier, Markowitz, Sharpe, and Black-Scholes all assume continuous change from one price to the next. Without that, their formulae simply do not work.

Alas, the assumption is false and so the math is wrong. Financial prices certainly jump, skip, and leap—up and down. In fact, I contend the capacity for jumps, or discontinuity, is the principal conceptual difference between economics and classical physics. In a perfect gas, as molecules collide and exchange heat, their billions of individually infinitesimal transactions collectively produce a genuine "average" temperature, around which smooth gradients lead up or down the scale. But in a financial market, the news that impels an investor can be minor or major. His buying power can be insignificant or market-moving. His decision can be based on an instantaneous change of heart, from bull to bear and back again. The result is a far-wilder distribution of price changes: not just price movements, but price dislocations. These are especially noticeable in our Information Age, with its instantaneous broadcasting by television, Internet, and trading-room screen. News of a terrorist attack in Indonesia flashes across the globe in seconds to millions of investors. They can act on it, not bit by bit in a progressive wave, as conventional theorists assume, but all at once, now and instantaneously. The effect can be exhilarating or heart-stopping, depending on whether you gain or lose.

It can also be embarrassing. Few things so panic investors as a sudden price drop, and the mutual fund industry sometimes goes to extraordinary lengths to "manage" emotions. In 2000 a Milwaukee mutual fund company, Heartland Advisors Inc., hit turbulence when the market value of some of its bond investments plummeted to $80 per $100 face value, from as high as $98. But that did not show up immediately in its daily price reports. Instead, according to the Securities and Exchange Commission, the fund's data supplier recorded a long, slow, and gentle decline over a period of weeks—at fifty cents a day. It did little good: When word eventually got out, Heartland investors stampeded to the exits. The price of one fund collapsed by nearly 70 percent in a single day. The SEC later sued the data supplier, which settled without admitting or denying the charges.

But discontinuity can be profitable, too. For more than a century, the New York Stock Exchange has had a system of "specialists." These are traders on the exchange floor who each specialize in the shares of a few companies, maintaining an order book, and, when the buys do not match the sells, stepping in with their own money to complete trades. Their function, according to the rules, is to "ensure the continuity of the market." Lately, they have come into disrepute in the post-bubble scandals that have engulfed most of Wall Street. In the SEC study of the 1997 collapse mentioned earlier, the agency found specialists in the most tumultuous twenty-four minutes were powerful net buyers; the volume of their purchases exceeded their sales by a ratio of 2.06. These were good bets: Prices did recover.

5. In Markets, Time Is Flexible.

If time is money, then the currency on Wall Street needs reform. Conventional financial analysis is a welter of conflicting views of time. One, implicit in conventional finance theory: Time is measured by the clocks and is the same for all investors. When calculating risk under the Capital Asset Pricing Model, the formulae assume all investors think and breathe very much alike, holding the securities in question for exactly the same length of time. The contradictory view, popular among market pundits: Time is different for every investor. Each time-scale you consider, each holding-period for a stock or bond, has its own kind of risks. Under this view, a quick day-trade poses entirely different scales of risk than does a six-month investment—and in most eyes, the day-trader is the more likely to go broke.

Things need not be so complicated. The genius of fractal analysis is that the same risk factors, the same formulae apply to a day as to a year, an hour as to a month. Only the magnitude differs, not the proportions. In fractal analysis, a price series is like a long, folding car antenna. You can look at its full length, segment by segment; or you can simply collapse it so each length is stacked inside the next. This is the scaling property of financial price series, as described earlier. Statistically speaking, the risks of a day are much like those of a week, a month, or a year. But the price variations scale with time.

Again, all charts look the same. In the case of cotton, I found all the price variations followed the same statistical properties for days over a few decades and for months over eighty years. All the lines were equally wiggly. Why would this be? First, I surmise, economics differs from physics in having no intrinsic time scales. The chart of a day's activity looks like that of a month because, from the narrow viewpoint of the probability of losses or gains, a day really is like a month. Yes, some time-scales have some meaning: Companies report their financial results quarterly and annually. A trading day has its own internal rhythm, as exchanges open and close in the procession of the day around the globe. But these are cyclical patterns, which financiers and economists have long since learned to make disappear statistically when building a model or investment strategy; that is the meaning of seasonal adjustment. And these differences are nothing like the immutable, fundamental differences in time-scale that arise in physics. There is, in finance, no barrier like that between the subatomic laws of quantum physics and the macroscopic laws of mechanics.

In fractal analysis, time is flexible. The multifractal model describes markets as deforming time—expanding it here, contracting it there. The more dramatic the price changes, the more the trading time-scale expands. The duller the price chart, the slower runs the market clock. Some researchers have tried linking this concept to trading volume: High volume equals fast trading time. That is a connection not yet established, and it need not be. Time deformation is a mathematical convenience, handy for analyzing the market; and it also happens to fit our subjective experience. Time does not run in a straight line, like the markings on a wooden ruler. It stretches and shrinks, as if the ruler were made of balloon rubber. This is true in daily life: We perk up during high drama, nod off when bored. Markets do the same.

6. Markets in All Places and Ages Work Alike.

If you throw a cat into the air, it will land on its feet. It is one of the little miracles of animal neurology with which we are all familiar. But even more miraculous: If the cat happens to brush against an obstacle on the way down, the edge of a table, for instance, its body in mid-air will spontaneously adjust course, to avoid a collision. How does it do it?

That was the subject of one of my stranger research collaborations, when I was for a year a visiting professor of physiology at Albert Einstein School of Medicine, in New York. My host, Professor Vahe Amassian, wanted to get to the bottom of this mystery, wiring a cat's brain to observe the pattern of neuronal firing in mid-flight. (Yes, it is a bit scary to see all those electrodes coming out of its head.) But I urged him to take it easy and first go back to basics: What does the cat's brain activity look like when nothing is happening? When it is sleeping? We must first understand the cat's brain at rest before we can understand its brain in action. So the young post-docs there stopped tossing the cats, and started stroking them. They pet them, set them to purring, and watched the instrument read-outs of the brain patterns. They let the cats sleep, and again watched the instruments. Amazing fact: The cats' brains were constantly abuzz with activity, even with little or no outside stimuli. They had a spontaneous firing of neurons. Or, as an economist would put it, in the absence of any "exogenous" input, "endogenous" activity continues, according to complex rules we started to unravel.

Now, it is not possible to observe a stock market in isolation from the world around it; but the principle remains. In a market there is, I believe, a spontaneous internal life, an inherent activity that comes from the way people come together, organize themselves in banks or brokerages, and exchange assets. This internal process does not make prices on its own; but it is certainly part of the price-setting mechanism—as much as the news, the bankruptcies, the economic reports, the wars and earnings announcements on which we more commonly focus our attention. It comprises the endogenous variables in the price-making equations, the cog ratios inside the black-box machinery that inputs an economic input and outputs an IBM stock price. To whatever extent one market is like another, you would expect this endogenous activity to be partly responsible for the similarity.

One of the surprising conclusions of fractal market analysis is the similarity of certain variables from one type of market to another. My cotton study found the same wild degree of price swings, over more than a century of trading records. I conjecture that the process generating American cotton prices changed only in scale, not in nature.

We mathematicians and physicists love what we call an invariance. That is a property that remains unchanged, no matter how you transform the data, shape, or object under study. Fractal geometry is the mathematics of one such invariance in the physical world—the study of patterns, in space or time, that remain the same even as the scale of observation changes. Statisticians have a kindred concept, called stationarity: A stationary time series has the same basic statistical properties throughout. Economists argue their field may be different. Economist Jacob Marshak once proclaimed at a meeting I attended that the only economic invariance he could imagine concerned the equality between the number of left and right shoes—and not even that could be trusted. Following that thinking, many recent models of price variation try to explain the obviously shifting pattern of volatility by inserting parameters that change by the day, hour, and second; such are in the GARCH family mentioned earlier. I am an optimist. I would rather not dismiss the existence of invariances but continually look for them hiding in non-obvious places. Invariances make life easier. If you can find some market properties that remain constant over time or place, you can build better, more useful models and make sounder financial decisions. My multifractal model works with just such a set of consistent parameters.

7. Markets Are Inherently Uncertain, and Bubbles Are Inevitable.

What does it feel like, to live through a fractal market? To explain, I like to put it in terms of a parable:

Once upon a time, there was a country called the Land of Ten Thousand Lakes. Its first and largest lake was a veritable sea 1,600 miles wide. The next biggest lake was 919 miles across; the third, 614; and so on down to the last and smallest at one mile across. An esteemed mathematician for the government, the Kingdom of Inference and Probable Value, noticed that the diameters scaled downwards according to a tidy, power-law formula.

Now, just beyond this peculiar land lay the Foggy Bottoms, a largely uninhabited country shrouded in dense, confusing mists and fogs through which one could barely see a mile. The

Kingdom resolved to chart its neighbor; and so the surveyors and cartographers set out. Soon, they arrived at a lake. The mists barred their sight of the far shore. How broad was it? Before embarking on it, should they provision for a day or a month? Like most people, they worked with what they knew: They assumed this new land was much like their own and that the size of lakes followed the same distribution. So, as they set off blindly in their boats, they assumed they had at least a mile to go and, on average, five miles.

But they rowed and rowed and found no shore. Five miles passed, and they recalculated the odds of how far they had to travel. Again, the probability suggested: five miles to go. So they rowed further—and still no shore in sight. They despaired. Had they embarked upon a sea, without enough provisions for the journey? Had the spirits of these fogs moved the shore?

An odd story, but one with a familiar ring, perhaps, to a professional stock trader. Consider: The lake diameters vary according to a power law, from largest to smallest. Once you have crossed five miles of water, odds are you have another five to go. If you are still afloat after ten miles, the odds remain the same: another ten miles to go. And so on. Of course, you will hit shore at some point; yet at any moment, the probability is stretched but otherwise unchanged.

It is a logical consequence of scaling. As I have stated often, the distribution of price changes in a financial market scales. Like the proportion of billionaires to millionaires in Pareto's income formula, so the proportion of big changes to small changes in a financial price series follows a consistent pattern—and it results in wilder price swings than you might otherwise expect. Rephrase this in the language of conditional probability: Given that event X has happened, what are the odds that Y will happen next? In Pareto's case, the scaling formula means that the odds of making more than ten billion once you make more than one billion are the same as those of making more than ten million once you make more than one million. With financial prices, scaling means that the odds of a massive price movement given a large one are akin to those of a large movement given a merely sizeable one. In both cases, the proportions are controlled by a scaling exponent, α.

A mind-bending paradox, to be sure. But to bring it down to earth, rewrite the parable and set it at the New York Stock Exchange. For explorers, read investors. For fogs, read the limits of our knowledge. And for the lakes, read the prices of 10,000 different securities. Have you alighted upon a stock the price of which will run and run until your profits are so vast you cannot count them? Or have you found a loser that, just as the price seems to take off, unexpectedly falls short? Are you living through a price bubble that will burst at any moment, so you should stay away? Or have the fundamental economic rules of the game changed, so that only a timid fool would not invest? Such is the confusion of scaling. It makes decisions difficult, prediction perilous, and bubbles a certainty.

8. Markets Are Deceptive.

Bubbles are dramatic—but the tendency of markets to deceive and confuse is an everyday affair. Consider chartists, who try to spot patterns in the market. The sophistication of these techniques varies greatly. Some are mere eyeball hunches: A pattern in an index or price chart looks like one that has happened before, and so you bet the chart will keep moving in the same way. Others are more elaborate. The best-known example is the Elliott Wave. Ralph Nelson Elliott was a Kansas-born accountant who spent much of his working life reorganizing railroads and state finances in Central America and who, during a debilitating illness, devised a new charting methodology. Investor psychology, he felt, moves in waves of optimism and pessimism; and these waves can be seen in the stock market again and again, at different times and at different time-scales. His theories gained attention in the 1930s, when he correctly predicted a few market turns, and public interest in them revived in the 1980s. But Wave prediction is a very uncertain business. It is an art to which the subjective judgment of the chartist matters more than the objective, replicable verdict of the numbers. The record of this, as of most technical analysis, is at best mixed.

People want to see patterns in the world. It is how we evolved. We descended from those primates who were best at spotting the telltale pattern of a predator in the forest, or of food in the savannah. So important is this skill that we apply it everywhere, warranted or not. We see patterns where there are none. Between the wars, Evgeny Slutzky, a Soviet statistician, showed how even the record of a Brownian motion—accumulation of a coin-toss game—can appear deliberate and ordered. The eye spontaneously decomposes it into up and down cycles, and then into smaller cycles that ride on the bigger cycles, and so on. Add more data, and more cycles appear. These are not real, of course. They are the mere juxtaposition of random changes.

How much more prone to spurious patterns, then, is an economic or financial price series? As described earlier, the long-range dependence in prices creates a kind of tendency in the data—not towards any particular price level, but towards price changes of a particular size or direction. The changes can be persistent, meaning that they reinforce each other; a trend once started tends to keep going. Or they can be anti-persistent, meaning they contradict each other; a trend once begun is likely to reverse itself. The persistent variety, especially those with an H exponent near 0.75, are especially curious, and these are the type common to many financial and economic data series. In our research in the late 1960s, Wallis and I generated such records by the purest operations of chance. Nevertheless, they all appeared to display a long, slow, up-down cycle of three; upon those long waves, smaller and more numerous cycles seemed to interpolate themselves. When we looked at a small section of the record, we again saw three waves, each a third shorter than the section.

One of the more controversial theories of the global economic cycle displays just such a rhythm of three. In 1925, Russian economist Nikolai Dmitrievich Kondratieff postulated the existence of "long waves" of growth and decline in the major economies of the West. Each wave averaged fifty-four years, the first beginning in the 1780s and, he forecast, the third to end in the 1940s. Since World War II, economists have debated whether a fourth cycle has begun or whether Kondratieff was simply mistaken. I cannot judge; but I do observe that, by the sheer operations of chance, one might easily see a three-wave, fifty-year pattern emerging over a century and a half of data. That we would ascribe economic meaning to it may tell us more about the way our minds work than about the way the levers of production and growth operate.

Indeed, so deceptive is long-term dependence that it has found a place in the toolkit of our age's ultimate fabulists: Hollywood. I have devised fractional Brownian motions "forgeries" that yield quick, realistic-looking landscapes. A demonstration follows. The illustra-

The deceptive power of chance. The Alps? The Mountains of the Moon? In fact, this is only the handicraft of a computer programmed by R.F. Voss. When it appeared in Mandelbrot 1982, this diagram defined the state of the art in computer graphics. There is no geophysics in this diagram, only suitable randomness and fractality. The point is the same as in fractal finance: Chance alone can produce deceptively convincing patterns.

tion looks, for all the world, like a relief map of the Himalayas. It is in fact the handiwork of a computer, running a simulation by the purest operations of an appropriate form of chance.

It takes no great leap of the imagination to see how such spurious patterns could also appear in otherwise random financial data. This is not to say that price charts are meaningless, or that prices all vary by the whim of luck. But it does say that, when examining price charts, we should guard against jumping to conclusions that the invisible hand of Adam Smith is somehow guiding them. It is a bold investor who would try to forecast a specific price level based solely on a pattern in the charts.

9. Forecasting Prices May Be Perilous, but You Can Estimate the Odds of Future Volatility.

All is not hopeless. Markets are turbulent, deceptive, prone to bubbles, infested by false trends. It may well be that you cannot forecast prices. But evaluating risk is another matter entirely.

Step back a moment. The classic Random Walk model makes three essential claims. First is the so-called martingale condition: that your best guess of tomorrow's price is today's price. Second is a declaration of independence: that tomorrow's price is independent of past prices. Third is a statement of normality: that all the price changes taken together, from small to large, vary in accordance with the mild, bell-curve distribution. In my view, that is two claims too many. The first, though not proven by the data, is at least not (much) contradicted by it; and it certainly helps, in an intuitive way, to explain why we so often guess the market wrong. But the others are simply false. The data overwhelmingly show that the magnitude of price changes depends on those of the past, and that the bell curve is a nonsense. Speaking mathematically, markets can exhibit dependence without correlation. The key to this paradox lies in the distinction between the size and the direction of price changes. Suppose that the direction is uncorrelated with the past: The fact that prices fell yesterday does not make them more likely to fall today. It remains possible for the absolute changes to be dependent: A 10 percent fall yesterday may well increase the odds of another 10 percent move today—but provide no advance way of telling whether it will be up or down. If so, the correlation vanishes, in spite of the strong dependence. Large price changes tend to be followed by more large changes, positive or negative. Small changes tend to be followed by more small changes. Volatility clusters.

What use is that? Plenty, if you are in the business of managing, avoiding, or profiting from risk. A bank is required, by its regulators, to estimate the value of its market assets daily, and set aside a certain amount of capital as a cushion against loss. A better, more-accurate way of estimating those potential losses would save the bank money and the financial system grief. A fund manager or investor who cannot tolerate the risk of a large loss might, when the financial storm signs are up, simply trim his sails and avoid bold bets. And options traders strive to profit from risk. They devise strategies and products—straddles, swaptions, barrier options—that pay best when they predict the future volatility best. They trade volatility; they even quote prices in "vols." The Chicago Board Options Exchange since 1993 has listed a product, the VIX, that is a bet on how volatile the S&P 500 will be in thirty days. As you would expect with so much money involved, the industry's analysts have devised many methods for forecasting the volatility— and (whether or not they say it) most recognize that the standard models do not work.

Of course, you cannot predict anything with precision. Forecasting volatility is like forecasting the weather. You can measure the intensity and path of a hurricane, and you can calculate the odds of its landing; but, as anyone who lives on the U.S. Eastern Seaboard knows, you cannot predict with confidence exactly where it will land and how much damage it will do. Nevertheless, work on such meteorological ideas has begun in finance. A first step is agreeing on a way to measure the intensity and path of a market crisis. The famous Richter Scale is the analogy most drawn upon. It measures the energy released by an earthquake on a logarithmic scale; for instance, a catastrophic quake of magnitude 7 packs ten times as much energy as a merely devastating quake of magnitude 6. What is a financial market's analog to energy? Volatility, some have surmised. Thus, two University of Paris researchers recently devised an Index of Market Shocks according to which there have been ten financial "quakes" since 1995. The Russian market crash of 1998 was a major tremor of 8.89 on the IMS scale. The biggest: the Twin Towers attack of September 2001, registering 13.42.

The next step is forecasting—but here, work is just beginning. Researchers in Zurich, working on their own scale for currency market crashes, found their index seemed to predict storms, albeit only over a short time-horizon. In the week of October 5–9, 1998, dollar/yen rates gyrated an extraordinary 15 percent. A few hours before the worst of the crisis, the researchers found, their index had soared from a level below 3 to one above 10. It "gave an early warning that the situation was very unstable," they reported.

You cannot beat the market, says the standard market doctrine. Granted. But you can sidestep its worst punches.

10. In Financial Markets, the Idea of "Value" Has Limited Value.

Value is a touchstone to most people. Financial analysts try to estimate it, as they study a company's books. They calculate a break-up value, a discounted cash-flow value, a market value. Economists try to model it, as they forecast growth. In classical currency models, they input the difference between U.S. and Euro zone inflation rates, growth rates, interest rates, and other variables to estimate an ideal "mean" value to which, over time, they believe the exchange rate will revert.

All this implies that value is somehow a single number that is a rational, solvable function of information. Given a certain set of information about an asset—a stock, a bond, or a pair of woolen *culottes*—everybody if equally well-placed to act will deduce it has a certain value; they will all hang the same price tag on it. Prices can fluctuate around that value; and it can be hard to calculate. But value, there is. It is a mean, an average, something certain in a chaos of conflicting information. People like the comfort of such thinking. There is something in the human condition that abhors uncertainty, unevenness, unpredictability. People like an average to hold onto, a target to aim at—even if it is a moving target.

But how useful is this concept, really? What is the value of a company? Well, you say, it is the price the market in its collective wisdom hangs on it. But how so? The most common index for market value is the price-earnings ratio, or P/E. Take Cisco Systems again, the supreme example of an Internet bubble stock. At its peak, the P/E reached a stratospheric 137. Put that into perspective. Any investor who actually believed that to be the company's intrinsic value would have had to assume its earnings would keep up the same torrid pace for at least another decade—by which point Cisco's market value would have exceeded the annual production of the entire U.S. economy. After the bubble burst, of course, the story changed. Cisco's P/E at the market nadir of early 2003 had fallen to 26. Oddly enough, by then its earnings growth was actually faster than in the bubble days: 35 percent. Does any of this make sense? Ah, you say, it was not the company's business fundamentals, but the market's appetite for technology companies that changed—and that is as much a part of the measure of intrinsic value as balance sheet or cash flow. Really? If that is so, then surely the "real" value of Cisco changes every month, every week, every day—even tick-by-tick on the stock exchange. And if that value changes constantly, then of what practical use is it to any investor or financial analyst weighing whether to buy or sell? What use is a valuation model with new parameters for every calculation?

Point taken, you say. Then value is, perhaps, some function of cost—the cost of producing a steel ingot, the cost of replacing a factory, the cost of buying a company's individual pieces, broken up. How so? What is the cost of Microsoft Office software? Easy, you say: Add up the latest development budget, overheads, finance charges, and operational expenses for the relevant Microsoft division. But how much should we include of the cost of earlier Office generations, products without which the latest Office would not exist? How about the cost of the Windows operating system, the basic software with which Office was designed to work? How about the cost of installing and maintaining Office on millions of customers' computers, without which Office would not have the "network economies" that have been so crucial to its growth? Such questions, difficult enough in a manufacturing economy, become intractable in our modern information economy, in which so much money changes hands for the mere right to use somebody else's intangible ideas. And even if we could agree on a cost, how could we ever derive a useful formula for translating it into a price? Things sell below cost all the time. The price of a dress can drop 90 percent, simply by moving it from the shop window at the start of the season to the basement clearance rack at the end of the season.

Point taken, you say. But intellectual property and financial assets are unusually insubstantial items. What about hard assets? Well, commodity prices are at least as wild as stock prices. Cotton prices flipped around so wildly you could not say that average or variance, the standard parameters of measurement, had much meaning. And what "real" value would you have assigned to silver in the winter of 1979–1980, when prices nearly trebled in the space of just six weeks? Property prices are no more substantial. As anyone buying or selling a house knows, "average" prices have no significance: The quoted survey figures are based on just a few sales scattered around a neighborhood and can apparently change by the day. And even those figures show bizarre patterns: In the late 1990s, London house prices more than doubled. So divorced from any idea of intrinsic value did property become that one developer rehabilitated a former public restroom, to sell as a small "cottage" for about £125,000—more than six times the average London wage.

To be sure, I do not argue there is no such thing as intrinsic value. It remains a popular notion, and one that I myself have used in some of my economic models. But the turbulent markets of the past few decades should have taught us, at the least, that value is a slippery concept, and one whose usefulness is vastly over-rated.

So how, you ask, does one survive in such an existentialist world, a world without absolutes? People do it rather well all the time. The prime mover in a financial market is not value or price, but price differences; not averaging, but arbitraging. People arbitrage between places or times. Between places: I had a friend who made his life as graduate student less tough by buying a convertible cheaply in his snowy home state, Minnesota, repairing it with his own hands, and then driving it to sunny California to sell dear. And arbitrage between times: A scalper buys a block of tickets today, and hopes to profit next month by reselling them dearly once the show is sold out. These arbitrage tactics assume no "intrinsic" value in the item being sold; they simply observe and forecast a difference in price, and try to profit from it. Of course, I am by no means the first to suggest the importance of arbitrage in financial theory; one of the latter-day "fixes" of orthodox finance, called Arbitrage Pricing Theory, tries to make the most of this. But a full understanding of multifractal markets begins with the realization that the mean is not golden.

CHAPTER XIII

In the Lab

IF YOU TAKE THE No. 4 streetcar from the center of Zurich, heading down the eastern lake shore, you will eventually come to the old Mill Museum, a four-story, century-old factory now housing worthy exhibitions on cereals, the food industry, and the age-old human cycle of famine and surplus, boom and bust. Next door, however, is a kind of laboratory for boom and bust—a test reactor, its founder calls it. "What we're doing is quantum theory for finance," says Richard Olsen.

His company, Oanda.com, looks like just another small financial house. Barely twenty-five people man its market-making screens, trade e-mail with customers, or work its computers. Its Web site, on foreign exchange markets, is good but, at first sight, nothing special. It has instant currency converters, live quotes, news, scholarly articles on market theory, trading games, downloadable software to analyze the market, and—now something out of the ordinary—a service that lets you bet real money on currency rates. If you open an account, you get what looks like a front-row seat at a Forex dealer's trading screen. On your PC, you can chart the dollar-yen or euro-sterling market, project future price movements, work out a trading strategy, and then place a bet, with real money. It can be as little as one dollar. Launched in 2001, the service in early 2004 had about 10,000 customers who had deposited money to trade. Most were amateurs, taking a flutter. But Oanda also attracts some big money. All told, its customers trade about $1 billion worth of yen, euro, dollar, bhat, or pesos a day.

It is, in short, a small-scale model of the real currency market. One problem with almost all economic or financial research is limited information. If you want to study a market, you can get lots of generic numbers—indices, price quotes, volume. If you are inside a brokerage house, you can supplement that with precise information about what your own clients are doing, and, to some extent, why. But you can never see what other firms' customers are doing. You can never get the whole picture, the satellite view. That is what Oanda.com provides Olsen and his handful of math and finance Ph.D.'s: the insight, both general and particular, of what people actually do in a market. They analyze it on their computers. They study customer behavior, how and when people open and close positions, how long they hold a position, what they do, and, to some extent, why.

"I have this terrible sense of frustration," says Olsen. "We send space shuttles into orbit; we send probes to Mars; but we haven't studied the financial markets. We literally know nothing about how economics works. I want to break that deadlock. I want to change financial markets into something as efficient as engineering."

I share his frustration. It is beyond belief that we know so little about how people get rich or poor, about how it is they come to dwell in comfort and health or die in penury and disease. Financial markets are the machines in which much of human welfare is decided; yet we know more about how our car engines work than about how our global financial system functions. We lurch from crisis to crisis. In a networked world, mayhem in one market spreads instantaneously to all others—and we have only the vaguest of notions how this happens, or how to regulate it. So limited is our knowledge that we resort, not to science, but to shamans. We place control of the world's largest economy in the hands of a few elderly men, the central bankers. We do not understand what they do or how, but we have blind faith that they can somehow induce the economic spirits to bring us financial sunshine and rain, and save us from financial frost and pestilence. If there is one message I would wish to survive this book, it is this: Finance must abandon its bad habits, and adopt a scientific method.

I do not claim to have the answers. I know a few things, gleaned by long research and free thought. Interest in my hypotheses has waxed and waned over the years. After an initial burst of trendiness in the 1960s, my then-half-formed financial theories fell out of fashion in the 1970s and 1980s. Only in the 1990s, by which time fractal geometry had become respectable and my theories had evolved, did a small but growing number of economists, mathematicians, and financiers join my labors. A rough estimate would place perhaps one hundred serious students of fractal financial and economic analysis around the world. Most are in academia, publishing in those few scholarly journals that allow such heresies into print. Some are in finance, trying to make money—though, in truth, our knowledge is still so limited that no one has yet to report great success.

I do not agree with all of these researchers or traders now experimenting with fractals—far from it, in some cases. The market trials, to date, do not count as scientifically sound endeavors, nor are they intended to be. They often mix a few of my fractal ideas with a large measure of other, often contradictory notions, in a spirit of "whatever works." Such are the exigencies of real-world finance— or academia! And, it must also be added, "fractal" has attained the status of a mini-cult. As in any cult, it has its share of opportunists.

But I applaud any serious efforts to push forward our understanding of how the financial system really works. And so, in concluding this book, I mention a few strands of current experiment, without endorsement or comment; the reader may draw his own conclusions. And I follow that with my own list of a few serious research questions that, in my view, need addressing.

"THIS IS like an atomic reactor," Richard Olsen enthuses about his project. "We can look inside and see how a financial market operates."

Olsen is a painfully earnest, lanky fifty-one-year-old with a manner more suited to scholar than trader. In the world of Forex, where he is well-known among the big bank research departments, he is viewed as something of a boffin: brainy, dedicated, and perhaps a bit eccentric. He got a master's in politics and economics from Oxford and a Ph.D. in law from Zurich and worked among the financiers of Zurich. But he quickly became a prophet for an important new faith in financial research: high-frequency data. A century ago, even yearly data on broad trends were hard to come by. Then reporting of monthly, weekly, and daily prices improved on exchanges and in newspapers. But the real data stream is tick-by-tick, quote-by-quote, transaction-by-transaction—and that was available only in a few places, such as on the New York Stock Exchange. So in the 1980s news services like Reuters began to see some value in transmitting instant-by-instant numbers to paying clients—and that is where Olsen and his colleagues in Zurich saw opportunity. They amassed, debugged, and began studying what has become one of the world's biggest databases of tick-by-tick foreign exchange quotations. For academics, it has been a boon; scores of scholarly finance articles have been published based upon it. But the big banks to whom Olsen also hoped to sell use of the database were not much interested. His firm was liquidated.

Oanda.com was his next idea for studying the market, which he founded in 1996 with a school friend and computer science professor, Michael Stumm. And it has been an entirely different story. In 2003, according to its reports to the U.S. Commodity Futures

Trading Commission, its net capital more than doubled to $4.1 million—a tidy profit. Olsen Investment Corp., a sister company, manages some relatively small sums—30 million euros at the end of 2003—for customers in the foreign exchange market. The funds have performed fairly well. In 2003, the best fund returned 21.05 percent, the worst, 3.15 percent, according to audited reports. The performance difference from one fund to the next arises mainly from how much risk, or leverage, each fund tolerates; as is common in Forex, the riskiest funds have done the best—so far. But the trading strategy for all the funds is the same, and follows Olsen's computerized, quasi-fractal models of the market. "The world is just fractal, and if you try to view fractal markets from a Euclidean perspective you just get it wrong from A to Z," he says.

To him, a financial transaction is like a small explosion. Conventional theory holds that prices change continuously, and that each investor is as unimportant as the next. Their trades are like the collisions of molecules in a gas chamber—millions of tiny energy exchanges. Nonsense, Olsen says. His tick-by-tick data show plainly that prices jump. Quotes stutter. And investors vary greatly in importance and impact on the market. A more accurate metaphor is the chamber in an internal combustion engine: millions of small and large explosions drive the car forward, as the sparkplugs fire and the pistons churn.

As he sees it, in a well-functioning market small investors behave much like big investors, and make profits that scale proportionately. Only the industry's unfair commission structure and other idiosyncrasies tilt the game. Likewise, short-term traders act much like long-term investors—again, with measurable scaling factors. He can see this, he says, in the computers tracking his Web service, FXTrade. There, fees are abolished and interest is compounded second-by-second; big and small investors are on an equal footing, as they place their currency bets. To keep the system real, Olsen is registered as a market-maker, like Citibank or the other behemoths that rule the real currency markets. Olsen's computers keep his own quotes in line with those of the big banks, and also buy or sell real currency contracts to manage his own risk. Like other market-makers, he earns money on the spread, or the difference between the rates he sets to buy and sell a currency. But to those using his system, all that is invisible: What they see is just a currency market, and they can trade in it as often as they like, with whatever strategy or investment they like.

"It's fantastic," says one day-trader in Waco, Texas, L. B. ("Just L. B.," he says, when asked his first name) Myers. He rises every morning at 1:45 A.M. Texas time to catch the opening of the foreign-exchange market in London, naps until New York markets open, and then continues like that, trading and napping, around the clock. He has, he says, racked up "seven-digit" profits since he discovered Olsen's site while surfing the Web in 2002. Then he had just lost money in stocks, and was looking for something new. Equal degrees of obsession are visible on the chatline that Oanda.com provides its customers. Traders in Hyderabad, India, swap tips with investors in London and Dayton, Ohio. "Any USD/JPY traders out there?" asks one trader in China; "need some comments about where the JPY is going!" Another trader, from South Africa, tries to bet on a euro rise—and then realizes he called it wrong. "I'm either being taught a lesson in humility, or the trends changed." Then he adds: "Yup...was a lesson."

Most have no knowledge of or interest in Olsen's fractal notions. But to explain then to the curious, he has devised a theory of what he calls "heterogeneous markets." Orthodox economics is all wrong, he says. People are not rational, and they do not all think alike. Some are quick-trigger speculators who pop in and out of the market hundreds of times a day. Some are corporate treasurers, deliberately buying or selling big contracts to fund a merger or hedge an export risk. Some are central bankers, who trade only occasionally, and at critical moments. Others are long-term investors who buy and hold for months or years. Each one, operating on his own time-scale, comes together at one moment of trading, like all of time compressing into an instant, or the entirety of a rainbow spectrum focusing onto one white point. That is where the multifractal analysis comes in, he says: It is a mathematical tool for decomposing the market into its different elements, and seeing how they inter-relate and interact. And it suggests some real-world trading strategies. Using his models, his computers look for moments when the short-term traders are moving opposite to the long-term investors—and then he bets that the imbalance will correct itself.

In the end, he says, his goal is to make the financial system work better and more safely. If the real market worked like FXTrade, costs would come down, liquidity would rise. "The world economy is like your body," he says. "Your heart pumps six liters of blood a minute, and so if you weigh eighty kilos it would take about fifteen minutes to pump your body's weight. By that analogy, the world foreign exchange market should be transacting $40 trillion every ten minutes. Today we do $1 trillion or so in twenty-four hours. My claim is the global economy is close to a heart attack."

ON PARIS'S broad Boulevard Haussmann, some serious money is at play. Jean-Philippe Bouchaud and some colleagues at Capital Fund Management were running two hedge funds with combined capital of $725 million as of the end of 2003. The funds engage in statistical arbitrage: They use mathematical models and computer horse-power to find what they think is incorrect pricing in the market, or other unstable patterns on which they can bet. The individual bets are small; but it is, for them, a game of large numbers. Many small profits can mount. In 2002, their biggest fund, Ventus, reported a stock-market gain of 28.1 percent, this, in a year when the market overall had fallen by a third. But it is also a game of chance: In 2003, they were less lucky with gains of just 3.32 percent. Their other fund, Discus, in the futures market, reported a 14.1 percent profit that year. "With statistical arbitrage, there are ups and downs," Bouchaud says with a shrug.

Their strategy is part multifractal, part many other things. They have devised some algorithms of their own, mostly secret, to identify potentially profitable situations in the market. Their models calculate what Bouchaud calls a "center of gravity" for individual stocks and the market overall; if a price rises or falls too far, they interpret it as a trading signal. Nothing multifractal there; in fact, while the math is far more complicated, the basic concept is an old one of betting on a stock reverting to its mean value.

But then Bouchaud uses some techniques he says derive from multifractal analysis. They help plan the trades, build the portfolios, and, most important, avoid risking too much money at a time. He and some colleagues published some information about it in a 1998 scholarly paper; they called it "tail chiseling": Under conventional portfolio theory, based on all the old assumptions of Brownian motion in prices, you build a portfolio by laboriously calculating how all the assets in a portfolio vary against each other; good diversification would mean some stocks zig when others zag. But Bouchaud's method takes it as given that prices exhibit long-term dependence, have fat tails, and scale by a power law. He focuses, then, only on the odds for a crash—sharp, catastrophic price drops. After all, it is not small declines that wipe an investor out, it is the crashes. So their scaling formula minimizes the odds of too many of the assets in a portfolio crashing at the same time. They used that to draw a "generalized efficiency frontier"—analogous to Markowitz's original portfolio technique—to help pick a portfolio that maximizes returns for a given amount of crash-protection. As the paper put it, "the frequency of very large, unpleasant losses is minimized for a certain level of return."

Thus, it is not just the stock-picking that is important, but also the risk-protection. For the latter, Bouchaud says, multifractal thinking is most useful.

.   .   .

FRACTALS IN finance have followed a winding path. In the early 1990s the topic became very faddish, just as Wall Street, recovering from the 1987 crash, was searching for new ideas. Several funds formed to experiment with chaos theory, borrowed from physics and math; as mentioned, chaotic systems often exhibit fractal behavior, but the two fields are intellectually distinct. One firm, the Prediction Co., was formed by some researchers at the Santa Fe Institute and, for a time, got much publicity. A Boston fund manager, Edgar E. Peters of PanAgora Asset Management, wrote two books on fractal market analysis—though, he now says, he does not actually use it in the management of PanAgora's funds. The firm's conservative clientele just was not interested, despite the publicity. But the research has moved on since then, and fractals are back in vogue among many in finance, although, it must be said, as with any fashion there is often more show than substance.

It is, in my view, premature to be hoping for serious gains from fractal finance. There is still too much we do not know. What follows is a brief summary of a few of the practical questions that need resolution. They give a flavor of what might become possible with fractal analysis—and, I hope, they will inspire others to undertake serious study in the field.

Problem 1: Analyzing Investments

Wall Street likes to keep score: The Dow, P/E, book-to-market, EBITDA. . . No matter their meaning; the numbers keep multiplying. They help spot a trend, compare investments, measure performance, set bonuses, calculate returns. When it comes to measuring risk, however, the industry's toolkit is surprisingly bare. The two most common tools are α, or volatility, and β, or the degree to which a stock's price changes correlate to those of the market overall. These two numbers are used again and again, the latter in portfolio construction and corporate finance, the former in virtually every kind of risk calculation under the sun. Of course, both numbers only have meaning if prices vary mildly by the bell curve, which they certainly do not, so applying them to a stock price is like using a hammer to cut a plank. But even if the math were right, the underlying premise is peculiar: How could it possibly be that one and the same probability distribution can describe all and every type of financial asset? Surely silver prices do not vary in the same way as Treasury bills? How could Amazon stock follow the same curve as pulp-paper futures?

Taxonomy is important. Finance today is in the primitive state of natural history three centuries ago. Its concepts and tools are limited, and so it frequently confounds species. If we could find new, more accurate ways of discriminating among investments, we would have a major discovery on our hands. As investors, we could pick stocks more easily. As money managers, we could design portfolios more carefully. As financiers, we could decide with greater certainty whether a new factory or merger meets the company's targets.

Many researchers have, in fits and starts, tried to find such a new set of tools in my work. The first such was α, the exponent that measures how wildly prices vary—how "fat" the tails of the price-change curve are. I found an α of 1.7 for cotton, suggesting strong variation. I found wheat was closer to 2, the bell-curve case, suggesting milder variation. My student, Fama, went on to find that different stocks seemed to exhibit different α values with, for instance, values near the Gaussian 2 for such industrial heavyweights as Alcoa, Standard Oil, and General Foods, and values nearer a Cauchy 1 for Westinghouse, United Aircraft, and American Tobacco. But it was also immediately apparent from Fama's work that the method used to calculate α was critical; when he used different methods he got different estimates for the same stock.

Equally frustrating, to date, has been the effort to fashion a risk yardstick from H, the exponent measuring the dependence of price changes upon past changes. For instance, one study of eighteen different dollar markets—against the yen, pound, and others—found a range of values from 0.53 to 0.63. While those are all above the random-walk value of 0.5, no clear pattern has yet emerged to explain why each currency has the value it does. As mentioned earlier, another researcher, Peters of PanAgora, reported in 1994 what appeared to be a complete, logical system of variation of H by asset type. High-tech stocks had high dependence and H values; stable utility shares had H values closer to those of a random walk. That meant the high-tech stocks were more volatile, as conventional analysis tells us. Peters went on to argue that, for an investor, that made them a better bet because their price trends could be more easily perceived. But, again, the methods used are very finicky. If you look across all the studies to date, you find a perplexing range of H values and no clear pattern among them. For instance, what is H for the dollar-Deutschemark market? Answer: Six different studies found six different values, between 0.55 and 0.64. Dollar-yen? An almost identical range of values. In the stock market, various authors estimated the S&P 500 index at somewhere between 0.53 and 0.74. The list could go on. No consensus.

There are, however, entirely different approaches to fractal analysis. At Yale some of my students have tried creating what you might call a fractal fingerprint of a stock. The idea is to use the record of an individual stock's price fluctuations to drive a repetitive, fractal process—rather like using the data from a particular patient's EKG readings to punch the paper roll of an old player piano. It sounds eccentric. But you could imagine that such a process would systematically highlight certain differences from one data series to another. For instance, certain troublesome heart rhythms in the raw EKG data might produce a piano roll with a characteristic pattern of soft high notes, or an absence of notes around middle "C"—each of them instantly recognizable. In the same way, certain patterns of price variation would produce a telltale pattern on the fractal fingerprint. As the diagrams below suggest, you can see some sharp differences in fingerprints for different kinds of stocks. The Citigroup fingerprint has a clear diagonal line from top left to bottom right, suggesting a pattern of price variation with lots of small, successive fluctuations up and down, as you would expect from a stable bank stock. By contrast, the Sonus Networks chart shows an opposite diagonal, from bottom left to top right; that suggests a pattern of wild swings up alternating with wild swings down—again, as you would expect from a very risky tech stock. Others have used this technique to analyze the relationship between the Chinese and Taiwanese stock markets, and between individual stocks and the market overall.

Clearly, fractal investment analysis has more questions than answers today, and that should be no surprise. The conventional tools of modern financial analysis have benefited from more than a half-century of development, by thousands of economists and financial analysts. But in fractal analysis, relatively few people have yet undertaken serious work—and that, in fits and starts. It is high time the work began in earnest.

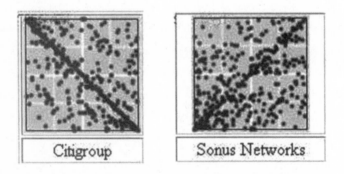

Fractal Fingerprints. One novel approach to fractal stock analysis is using the data on a stock's price variations to drive a simple, repetitive fractal process—thereby producing a unique graphical representation of how each stock's prices vary. At left is such a fingerprint for stable Citigroup, at right for risky Sonus Networks. As an aid to understanding, this technique confirms our intuition that the two companies' prices behave differently. But as a tool of financial analysis, it needs more work.

Problem 2: Building Portfolios

You can have your cake and eat it: Such is the underlying message of modern portfolio theory. It is an elaborate mathematical machinery for reducing risk without sacrificing too much profit. As elaborated by Sharpe in the Capital Asset Pricing Model described earlier, it starts with the premise that the expected profit from any security is the sum of two simple items. First is the return the stock earns simply by rising with the market overall, and second is whatever return it earns by marching to its own drummer. How much the stock rises or falls with the broad market index is measured by β, and man-centuries of time have been squandered by financial analysts calculating and studying this parameter. Generally speaking, a stock with a β of 1 moves in lockstep with the market overall. A stock with a higher β is hypersensitive to market moves; it magnifies the market risk and so, to bother buying it, you have to believe it is such a powerful growth stock that it is worth the risk. A stock with lower β is insensitive to market moves; it damps risk, and so may be more attractive in your portfolio even though you do not expect its price to rise much. With these assumptions, you can select stocks that mix and match risk and return and calculate, quite precisely, an optimal portfolio. Such is the theory. But practice often differs: Many fund managers have their own, peculiar styles of picking investments, and use the cold-blooded math of modern portfolio theory as a guide, a back-check that their picks are not piling on more risk than they thought.

But whether guide or master, modern portfolio theory bases everything on the conventional market assumptions that prices vary mildly, independently, and smoothly from one moment to the next. If those assumptions are wrong, everything falls apart: Rather than a carefully tuned profit engine, your portfolio may actually be a dangerous, careering rattletrap.

This was spelled out first by Fama. Conventional wisdom holds that, if you do the picks correctly, about thirty different stocks can provide an optimal portfolio. In fact, he found in a 1965 study, if you assume wild price variation you need many more stocks than that—perhaps three or four times as many. The wild swings of real markets mean you have to build in a wider margin of safety than conventional theory holds. In 2000, some researchers in France took his calculations to more detail. They found that, for nine stocks they studied on the Paris Bourse, the conventional methods consistently understated the basic market parameter, β. For instance, the standard method estimated French hotelier ACCOR to have a β of 0.91—meaning it is a good defensive stock to add to a portfolio. But when they re-calculated the number using a more realistic model of price variation, they found ACCOR's real β was 8 percent higher, or 0.98—meaning it is about as risky as the market overall. On average, the conventional methods underestimated β by 6 percent, they found. The implication: When you pick a stock by the conventional method, you may actually be adding risk rather than reducing it.

Can we build a new, correct portfolio theory? Not yet clear. Whether you use a conventional β or some new estimate of "real" β, the entire theory is founded on the belief that the market averages are important—that you can use the Dow or the CAC-40 as a good yardstick to measure the risk of every individual stock. But of what use is an average when the individual stocks diverge so widely and unpredictably from it? What is the "average" location of all the stars in the galaxy? A new approach is needed. Today, building a portfolio by the book is a game of statistics rather than intelligence: You start by assuming the market has correctly priced each stock, and so your task is simply to combine the particular stocks in your portfolio in such a way as to meet your investment goals. This is much like a painter taking the colors straight out of the tube, as mixed and labeled by the factory. But if the colors do not come pre-mixed, then the painter's eye for hue, intensity, and balance becomes more important. Likewise, if the stocks do not come pre-priced, if whatever drives the price-setting process is more complicated than expected, then the investment manager's skill at spotting good opportunities becomes more important. Indeed, in a non-Gaussian world, the investment manager might actually have to earn his high fees.

So what is to be done? For starters, portfolio managers can more frequently resort to what is called stress-testing. It means letting a computer simulate everything that could possibly go wrong, and seeing if any of the possible outcomes seem so unbearable that you want to rethink the whole strategy. The technology is called a Monte Carlo simulation. You tell a computer how you think prices vary—specifically, what kind of random-number generator it should use. You feed it all the initial data: the particular stocks, their price histories, your strategy for buying them. Then you press the start button. Using the rules of randomness you gave it, the computer starts generating a series of hypothetical prices for each stock—in essence, it simulates one investor's possible experience with the portfolio. Then it does it again, and again, thousands of times, like someone flipping a coin over and over to see if the odds for getting heads really are fifty-fifty. At the end, it totes up all the scores from all the runs: That tells you which simulated outcomes happened most often, and hence, which might be most likely in real life. It also tells you which outcomes are unlikely but, if they occurred, devastating. Finally, you use your own intelligence to decide whether you like the scenario the computer paints. If not, you decide the portfolio is too risky and you start again.

It sounds like a computational nightmare. Indeed, when this technique first appeared some decades ago in physics, it was not for the mathematically faint of heart. But computers are faster and cheaper now; software to perform these calculations now comes shrink-wrapped. You can simulate the performance of an options contract, for instance, in less than a minute on a standard personal computer. And so the technique has already spread over the past decade into many corners of finance. I urge that it become a standard tool of portfolio construction.

Problem 3: Valuing Options

What is an option worth? It depends on how you measure it.

One 2003 study, for the U.S. Financial Executives Research Foundation, compared six common ways of valuing stock options. By one method, it figured, a particular stock option it studied was worth $8.76 a share to the executive who received it from his company. But by another method, the thirty-year-old Black-Scholes equation, the same stock option was worth $25.27 a share. Which was right? Probably neither of them. Other studies have found even wilder errors. In the foreign exchange market, where $15 trillion of options were traded in 2001, one study found some dollar-yen options underpriced by 84 percent, and some Swiss franc–dollar options undervalued by 40 percent.

Valuing options correctly is a high-roller game, but the rules are all messed up. As described earlier, the most widely known formula was published in 1973 by Fischer Black and Myron Scholes, and it has been known for years that it is simply wrong. It makes unrealistic assumptions. It asserts that prices vary by the bell curve; volatility does not change through the life of the option; prices do not jump; taxes and commissions do not exist; and so on. Of course, these are simplifications to make the math easier. And so easy was it that, for the first fifteen years after its discovery, it was used blindly throughout options markets; it was viewed as a kind of financial alchemy that turned everything to gold. It let corporations hang a price tag on the stock options they granted their executives. It let banks devise new and ever-fancier financial products. It even allowed "portfolio insurance," a precisely calculated number of options designed to rise in value if your main stock portfolio falls. It seemed to be financial engineering of the highest form. It had abolished risk. Of course, the truth was re-discovered on Black Monday, October 19, 1987, when a sudden drop in stock prices was turned into a rout by a wall of insurance options crashing down on the market.

A fundamental problem is the Black-Scholes assumption of constant volatility—in essence, that the world does not change. Normally, to calculate an options price, you plug in a few numbers, including your estimate of how much the underlying stock price or currency rate fluctuated in the past; the suggested price falls out the back end of the formula. But if you run the equation in reverse, plugging real market prices into its back and pulling from its front the volatility that those prices would imply, you get a nonsense: a range of different volatility forecasts for the same options. A graphic example is below. It shows the implied volatility for several different flavors—different maturities and different strike prices—of the same kind of option. If Black-Scholes were right, this would be a

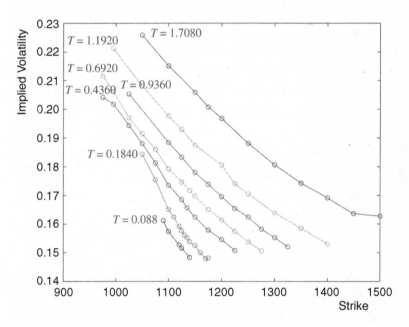

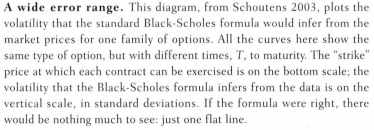

A wide error range. This diagram, from Schoutens 2003, plots the volatility that the standard Black-Scholes formula would infer from the market prices for one family of options. All the curves here show the same type of option, but with different times, *T*, to maturity. The "strike" price at which each contract can be exercised is on the bottom scale; the volatility that the Black-Scholes formula infers from the data is on the vertical scale, in standard deviations. If the formula were right, there would be nothing much to see: just one flat line.

very boring picture, one flat line for all the varieties. Instead, you see a whole range of errors, wandering across the chart. Indeed, the mistakes have a Rococo structure of their own, worthy years of study. In the options industry, where mistakes can cost millions, that is exactly what they have received. Hundreds of scholarly papers, several textbooks, and scores of financial conferences have been devoted to studying the errors.

Improving or replacing Black-Scholes is one of the liveliest sub-disciplines in mathematical finance. The most common approach is to try merely fixing the old formula. Software to correct the "volatility smile," the U-shaped pattern that Black-Scholes volatility errors often trace on graph paper, is now standard. Many adopt the GARCH methods mentioned earlier; while these produce better results than Black-Scholes alone, they are still not accurate. Some approaches mix ideas similar to mine with those of others. For instance, Morgan Stanley has used what is called a "variance gamma process" to value its own options books at the end of each trading day. This method, developed by Dilip B. Madan of the University of Maryland and two others, is a two-step formula. It starts with an equation to deform time, to make it jump ahead randomly before slowing again. It follows with a type of Brownian motion to generate a price. There are many others—and so far, no consensus in the industry about which work best. In the absence of clear answers, it has become a case of every man for himself. Even in the same firm, you can have one group using experimental new methods to price "exotic" options, a range of complicated, and highly profitable, products that banks devise for their corporate clients with special risk problems. You can have the compliance officers, responsible for making sure the bank does not lose too much money, using a modified Black-Scholes formula. And then you can have the traders themselves using all or none of the above, as their whim or experience dictate.

That is, I submit, no way to run the options business. Even if Wall Street is content, Main Street is not. In 2004, the main

American accounting body, the Financial Accounting Standards Board, revised the rules by which corporations account for the stock options they grant their executives. After the bursting of the Internet bubble, the obscene spectacle of greedy CEOs cashing in their options ahead of other shareholders stirred a political backlash. Upshot: FASB, under prodding from Washington, is requiring many companies for the first time to record their options as an expense—in other words, an employment cost that will hit their reported profits. That position has enraged many corporate chieftains, especially in the high-tech sector. Of course, they fear expensing options in any form will make them unattractive. But they also complain that there are no good valuation formulae.

"Despite results that are inherently inaccurate and unreliable for this purpose," groused Intel CEO Craig Barrett recently, "Black-Scholes is the only method available." He continued:

> If the standard-setters who support stock option expensing were required to certify their work, I wonder whether their tolerance for inaccuracy would be the same? I know of no situation where it would be acceptable for a CEO to certify that a company's results were 'kind of right'—the term used by FASB's Mr. Herz to describe the results produced by the Black-Scholes model....
>
> I support . . . corporate reform, but with all due respect, results that are 'kind of right' aren't good enough.
>
> Wall Street Journal, *April 24, 2003*

Problem 4: Managing Risk

By any measure, the late 1990s were a time of extraordinary growth and prosperity in much of the world—and yet, the global financial system still managed to lurch its way through six crises. The U.S. treasury secretary for part of that time, Lawrence H. Summers, counted them: Mexico in 1995; Thailand, Indonesia, and South Korea in 1997–1998; Russia in 1998; and Brazil in 1998–1999. The Indonesian crisis was especially severe: The country's quarterly real GDP plummeted 18.9 percent and its currency fell into a hole 526 percent deep. Each of these end-of-millennium upheavals spread from its origin to most parts of the globe, destabilizing currencies, knocking gaping holes in bank balance sheets, and, in many cases, causing a wave of bankruptcies. The fact that each country recovered and the global economy roared on again is a testament, not to good financial management, but to good luck.

So risk-management is now a hot topic among financiers and politicians. To safeguard against bankruptcy, most banks in the world are obliged by law to keep a certain amount of cash on hand—a capital reserve. It can be tapped *in extremis*, but its main purpose is to assure the rest of the world that all is safe, and the bank that has it is a safe partner with which to do business. That presupposes the reserve is large enough, and there lies the heart of the problem. In Basel, the Bank for International Settlements helps set the global standards for how much is enough, and since 2001 the world's bankers and finance ministers have been arguing over new rules. The old methods are inadequate, they agree. So what should replace them?

One of the standard methods relies on—guess what?—Brownian motion. The same false assumptions that underestimate stock-market risk, mis-price options, build bad portfolios, and generally misconstrue the financial world are also built into the standard risk software used by many of the world's banks. The method is called Value at Risk, or VAR, and it works like this. You start off by deciding how "safe" you need to be. Say you set a 95 percent confidence level. That means you want to structure your bank's investments so there is, by your models, a 95 percent probability that the losses will stay below the danger point, and only a 5 percent chance they will break through it. To use an example suggested by some Citigroup analysts, suppose you want to check the risk of your euro-

dollar positions. With a few strokes on your PC keyboard, you calculate the volatility of the euro-dollar market, assuming the price
changes follow the bell curve. Let us say volatility is 10 percent.
Then, with a few more strokes, you get your answer: There is only a
5 percent chance that your portfolio will fall by more than 12 percent. Forget about it.

The flaw should be obvious by now. The potential loss is actually
far, far greater than 12 percent. The problem is not merely that the
bell curve leads us to underestimate the volatility. That would be
bad enough, as it would understate the odds of loss. The problem is
worse than that. Assume the market cracks and you land in the
unlucky 5 percent portion of the probability curve: How much do
you lose? Well, 12 percent, you say. Wrong. Even the VAR model
recognizes that the actual loss could be greater; the amount beyond
the theoretical 12 percent is the "overhang." With a bell-curve
assumption, the overhang is negligible. But if price-changes scale,
the overhang can be catastrophic. As described before, once you are
riding out on the far ends of a scaling probability curve, the journey
gets very rough. There is no limit to how bad it could get for the
bank. Its own bankruptcy is the least of the worries; it will default
on its obligations to other banks—and so the final damage could be
greater than its own capital. That was the lesson from each international crisis, as losses spread from one interlinked financial house to
another. Only forceful action by the regulators put a firewall around
the sickest firms, to stop the crisis spreading too far.

Fortunately, bankers and regulators now realize the system is
flawed. So the world's central banks have been pushing for more
sophisticated risk models. One gaining popularity, based on something called Extreme Value Theory and borrowed from the insurance industry, is on the right track: It assumes prices vary wildly,
with "fat tails" that scale. But it does not commonly take account of
a further source of risk I have been describing: long-term dependence, or the tendency of bad news to come in flocks. A bank that
weathers one crisis may not survive a second or a third. I thus urge

the regulators, now drafting a New Basel Capital Accord to regulate global bank reserves, to encourage the study and adoption of yet more-realistic risk models. If they do not, Summers's list of six crises will just keep growing.

It is gratifying to find I am no longer alone on this point. After his trading house, LTCM, crashed in the 1998 Russian crisis, Myron Scholes wrote:

> Now is the time to encourage the BIS and other regulatory bodies to support studies on stress-test and concentration methodologies. Planning for crises is more important than VAR analysis.
>
> American Economic Review, *May 2000*

Aux Armes!

I am a persistent man. Once I decide something, I hold to it with extraordinary tenacity. I pushed and pushed to develop my ideas of scaling, power laws, fractality, and multifractality. I pushed and pushed to get out into the scholarly world with my message of wild chance, fat tails, long-term dependence, concentration, and discontinuity. Now I am pushing and pushing again, to get these ideas out into a broader marketplace where they may finally do some concrete good for the world.

Of course, I have my hypotheses about market dynamics; and I believe they are well founded. Others have opposing views. Even the most cursory trawl through the economics literature will find a perplexing cacophony of conflicting opinions—and, more invidious, contradictory "facts." Consider one example. Proposition: Prices are dependent over a time-span that is (a) a day, (b) a quarter, (c) three years, (d) an infinite span, or (e) none of the above. Which is the right answer? All of them, apparently, if you are to believe the conflicting economics literature. All these views you will find asserted as an unassailable fact in countless articles reviewed by countless worthy peers, and supported by countless computer runs, probability tables, and analytical charts. Wassily Leontief, a Harvard economist and 1973 Nobel winner, once observed: "In no field of empirical enquiry has so massive and sophisticated a statistical machinery been used with such indifferent results."

It is time to change that. As a first step, I issue a challenge to Alan Greenspan, Eliot Spitzer, and William Donaldson—Federal Reserve chairman, New York attorney general, and SEC chairman, respectively. In the April 2003 settlement of post-bubble fraud charges, the biggest Wall Street firms agreed to cough up $432.5 million to fund "independent" research. Spitzer's office amply documented that what passed for investment research before was not only wrong, but fraudulent. Since then, a long line of media and ratings firms have lined up to collect some of the loot to launch independent research businesses. But there has been precious little discussion of what, exactly, these researchers should research.

I suggest just a small fraction of that sum—say, 5 percent, in honor of the VAR analysis discussed above—be set aside for fundamental research in financial markets. Let the vast bulk of the money go where it usually does: ephemeral and contradictory opinions on which stocks to buy, which to sell, and whether to buy or sell at all. But let at least a widow's mite go to understanding how stocks behave in the first place. Let the Wall Street settlement help to fund an international commission for systematic, rigorous, and replicable research into market dynamics. Of course, $20 million is not enough; even if computers and doctoral students are cheap, proprietary data sources are not. But with that starting sum and wise leadership, such a commission would quickly draw contributions and investments from others, magnifying its impact.

A well-managed corporation devotes some portion of its research and development budget to fundamental research in fields of science that underlie its main businesses. Is not understanding the market at least as important to the economy as understanding solid-state physics is to IBM? If we can map the human genome, why can we not map how a man loses his livelihood? If millions, on the Internet, can contribute a few cycles of their home PCs to searching for a signal from outer space, why can they not join a coordinated search for patterns in financial markets?

On the night of February 1, 1953, a very bad storm lashed the Dutch coast. It broke the famous sea dikes, the country's ancient and proud bulwark against disaster. More than 1,800 died. Dutch hydrologists found the flooding had pushed the benchmark water-level indicators, in Amsterdam, to 3.85 meters over the average level. Seemingly impossible. The dikes had been thought to be safe enough from such a calamity; the conventional odds of so high a flood were thought to have been less than one in ten thousand. And yet, further research showed, an even greater inundation of four meters had been recorded only a few centuries earlier, in 1570. Naturally, the pragmatic Dutch did not waste time arguing about the math. They cleaned up the damage and rebuilt the dikes higher and stronger.

Such pragmatism is needed in financial theory. It is the Hippocratic Oath to "do no harm." In finance, I believe the conventional models and their more recent "fixes" violate that oath. They are not merely wrong; they are dangerously wrong. They are like a shipbuilder who assumes that gales are rare and hurricanes myth; so he builds his vessel for speed, capacity, and comfort—giving little thought to stability and strength. To launch such a ship across the ocean in typhoon season is to do serious harm. Like the weather, markets are turbulent. We must learn to recognize that, and better cope.

Notes

Fractal finance, in its full detail, is a beautiful and highly mathematical topic. We have avoided the use of equations so far in this book. But for the curious reader, in these Notes we provide a brief elaboration—mathematical and historical. Further detail can be found in this book's bibliography and http://www.misbehaviorofmarkets.com and in many cases more directly in http://classes.yale.edu/fractals/index.html.

Prelude
Introducing a Maverick in Science

xiv *"Paul H. Cootner..."*
From Cootner 1964.

xv *"The grand aim of all science..."*
An oft-repeated quotation of Albert Einstein, from *Life* magazine, January 9, 1950.

xvii *"Mandelbrot's life story..."*
All accounts of Mandelbrot's life in this book are based primarily on conversations between the authors, supplemented by Mandelbrot's own writings. A summary of his life and work may be found in Gleick 1987. An autobiographical essay plus additional biographical and bibliographical information is available at Mandelbrot's web site, http://www.math.yale.edu/mandelbrot. Separately, Mandelbrot is writing his memoirs.

xxiii *"And finally, he built..."*

It may be helpful to reference here his most significant papers in finance. The first on "fat tails" and scaling in finance were Mandelbrot 1962b and c, 1963a and 1967. The first on scaling and renormalization was Mandelbrot 1963c. The first on "long-term dependence" and the Hurst Effect were Mandelbrot 1965a and Mandelbrot and Van Ness 1968. The first on financial "bubbles" was Mandelbrot 1966a; on "trading time" and "subordination," Mandelbrot and Taylor 1967; on "multifractals," pointing out their possible applications to economics, Mandelbrot 1972.

xxiii *"That record, alone..."*

A summary of Mandelbrot's most recent work in finance may be found in a four-part series that appeared in *Quantitative Finance* (Mandelbrot 2001a-d). His most important past writings are being republished with extensive comments. So far, four volumes have appeared: Mandelbrot 1997a, 1999a, cover the topics suggested by their titles. The title of Mandelbrot 2002 is less descriptive therefore the contents of several chapters deserve to be singled out. Chapter H0, an overview of fractals and multifractals, is of wide general interest. Chapter H1, a close-up on a versatile family of cartoons, directly complements many topics discussed in this book. Chapter H5 describes the subtle path towards the exponent *H* that characterizes long dependence. Chapter H30 deals with the delicate problems that long dependences raises in the context of economics and finance. Chapters H11 to H14 and H27 reproduce often quoted papers Mandelbrot coauthored with J. Van Ness and J.R. Wallis. Chapters H21 to H24 reproduce his first step towards "cartoons."

The fourth *Selecta* volume, Mandelbrot 2004, tells nothing about finance but a great deal about the author, through richly illustrated stories, published for the first time, of how he discovered the Mandelbrot set.

The website www.math.yale.edu/mandelbrot promises to contain many of his articles. This is a work-in-progress but worth checking.

Chapter I
Risk, Ruin, and Reward

3   *"It might, said another, 'take a lifetime..."*

This account of the Russian crisis is based largely on contemporaneous news coverage by the *Wall Street Journal*, for which the newspaper won a Pulitzer Prize. Some of the key articles quoted are:

"Investors Find Few Havens from Russia—Most 'International' Funds Have Some Exposure to Former Soviet Union." By Charles Gasparino and Pui-Wing Tam. Aug. 28, 1998.

"Abreast of the Market: Russian Worries Overwhelm Stock Markets —U.S. Shares Decline 4.2%; Profit Fears Hurt Stalwart Names." By Robert O'Brien. Aug. 28, 1998.

"Down Market: U.S. Shares Plummet 512 Points, Bringing Bear Market Closer—A Drumbeat of Bad News Triggers Deep Pessimism Among Global Investors—Europe Faces a Grim Day." Wall Street Journal Europe Roundup. Sept. 1, 1998.

4   *"In the language of statistics ... "*
The crash probabilities were calculated from the daily differences in the natural logarithms of the Dow from 1916 to 2003. This is, actually, a simple task you can do yourself, with a spreadsheet program like Excel. Doing so demonstrates just how far the standard financial models deviate from common sense.

Download the index numbers from, for instance, Dow Jones & Co., at http://www.dowjones.com. Then use the spreadsheet to take the logarithm of each daily index number. Subtract each day's log from that of the day following to get the magnitude of the daily price changes; it is the change, not the index level itself, under study here. Now assume the changes fit the bell curve as the standard models suggest; so apply the cookbook formula for calculating the sample variance, s^2, of a Gaussian random variable:

$$s^2 = \frac{\sum_{i=1}^{n} (x_i - \bar{x})^2}{n-1}$$

Or in English: Sum all the squares of the difference between each daily change and the average change, and divide it all by the number of days minus one. (A reminder on the notation: The Greek capital letter, Σ, stands for taking the sum of the first, second, third, and all the other i terms of a series, up to the nth member. A bar over a variable means the average value.) To get the standard deviation, s, take the square root of all that. Now you know how much the index level "typically" varies from one day to the next; by the common math, 68 percent of all the variations will be within one standard deviation of the average. Next step: Calculate how "atypical" each crash day was. For that, you want to know how many standard deviations from the average value each crash was or, in the equation below, z:

$$z = \frac{x_i - \bar{x}}{s}$$

Lastly, with such a "z-score" for each crash, you can estimate the odds of such an event occurring if the standard Gaussian model were true. Many statistics textbooks publish the probabilities in tables; or you can find on the Web formulae to calculate them. Simply plug in the z-score, and crank out the odds.

The odds cited in the text are based on the standard deviation of the data set over the entire eighty-eight-year period (that is, n in the equation above is 88 years times 250 trading days/year). But similar results are found using a rolling 250-day standard deviation. A shorter, thirty-day calculation period makes some crashes seem slightly more probable—but only slightly so.

14   *"Apparently, a reluctance to recant..."*
From Buffett 1988.

15   *"For instance, the 'Black-Scholes'"*
After the Internet bubble, the U.S. Financial Accounting Standards Board began pushing American businesses to treat their executives' stock options as expenses. The question then arose: How to value the options? After hearings on the subject, in early 2004 the board was poised to broaden the list of approved valuation methods beyond the ubiquitous Black-Scholes formula. Further information is at http://www.fasb.org.

Chapter II
By the Toss of a Coin or the Flight of an Arrow?

26   *"As a graduate student at the University of Paris..."*
The thesis was "Contribution à la théorie mathématique des jeux de communication." Mandelbrot 1953.

30   *"One of the founders of modern probability theory..."*
From Gnedenko and Kolmogorov 1954.

34   *"It was in astronomy..."*
More of this tale is recounted in pp. 74-75 of Hall's 1970 biography of

Gauss. An especially entertaining account of Gauss's life can be found in Bell 1937: *Men of Mathematics*. This collection of mathematical biographies is inspiring, though not historically exact in every detail. The author was a professor of mathematics at California Institute of Technology, where I came to know him as a graduate student there.

34   *"Then do it again, and again..."*
The method of ordinary least squares is now standard in any elementary statistics course—albeit reduced to a cookbook form that Gauss and Legendre might barely recognize. If you assume that the measurement errors in your data—say, the results from a clinical trial of the efficacy of a new medicine at different doses—are Gaussian, then you can use a good hand calculator to crank out the "real" relationship between a given dosage, x, and a given therapeutic effect, y. A typical formula for it:

$$\hat{y} = \hat{\beta}_0 + \hat{\beta}_1 x_i + \varepsilon_i$$

The first term on the right $\hat{\beta}_0$ is the value of the therapeutic effect, $\hat{y}$, when the independent variable x, the dosage, is near zero—that is, when there is no medicine administered. The second term shows how quickly the efficacy rises as the dosage increases. And the third term is the error in each measurement. (The "hat" sign over a variable denotes an estimate from real data.) The result is, of course, an equation for a straight line—hence, the common lab-bench name for it is linear regression. It is, in essence, drawing a straight line through the cloud of data points to show an underlying "average" trend—if one actually exists. As will be seen, this method is used with reckless abandon in financial analysis. The key parameters, $\hat{\beta}_0$ and $\hat{\beta}_1$, in the regression equation are easily calculated with a pocket calculator and a standard formula.

36   *"But each has the same mathematical formula..."*
The formula for the bell curve is a somewhat forbidding function of one of those powerful mathematical constants that crop up in the most unlikely places. This constant, e, is an irrational number, with infinitely non-recurring digits, that starts as 2.7182. Its origin traces to the early seventeenth century, when it was found to be a useful part of the equations for calculating continuously compounded interest in finance. The equation for the bell curve gives the probability of an event—a particular level of IQ or human height, for instance—occurring in a given population that fulfils some basic conditions.

$$f(x) = \frac{1}{\sigma\sqrt{2\pi}} e^{-\frac{((x-\mu)/\sigma)^2}{2}}$$

Here, x is the particular level of the variable being studied: IQ of 110, say, or height of 6 feet 2 inches. The Greek letter μ, or mu, denotes the average value of all the x's in the population; and the Greek letter σ, or sigma, denotes the standard deviation—the benchmark of how broadly scattered all the x's are around the average. The "reduced" Gaussian corresponds to $\mu = 0$ and $\sigma = 1$. Thus, the particular value of x you are investigating determines where on the bell curve you are. If close to the average, the probability will be quite high; if on the far edges—the "tails"—the probability will be quite low. The standard deviation determines what kind of bell curve it is: whether squat and low with a high sigma, or narrow and tall with a low one. This is what makes the bell curve so popular: Just two numbers, the average and standard deviation, tell you all you need to know about a population, if it is Gaussian. And the bristling equation has long since been reduced to a simple function on good calculators, or an automatic formula in a spreadsheet.

38   *"They have infinite expectation..."*
Ironically, compared to the reduced Gaussian formula, the equation for a reduced Cauchy probability density is much simpler:

$$f(x) = \frac{1}{\pi(1+x^2)}.$$

Its graph as seen in the text, too, is a kind of bell curve—but with tails that flare out much farther and fatter.

Chapter III
Bachelier and His Legacy

44   *"Before Poincaré on that day..."*
The account of Bachelier's life and work given here derives from several sources. The first biographical sketch appears on pp. 172-177 of Mandelbrot 1975, expanded on pp. 392-395 of Mandelbrot 1982. It includes excerpts from Poincaré's report. The centenary in 2000 of Bachelier's thesis generated a virtual library of information. An excellent

selection of primary documents—correspondence, government records, and some of Bachelier's own papers—was published on the Web by Bachelier's last employer, the Université de Besançon, now called Université de Franche-Comté. They can be found at http://sjepg.univ-fcomte.fr/La_recherche/Libre/bachelier/page01/-page01.htm. There are other useful sources, such as Taqqu 2001 and Courtault 2000. An English translation of Bachelier's thesis is found in Cootner 1964.

50 *"Now, there had been a few..."*
This reference is Regnault 1863. The key observation, as translated in Taqqu 2001 runs as follow:

> There exists therefore a mathematical law which regulates the variations and the mean deviation of stock market prices, and this law, which seems never to have been noticed, is given here for the first time:
> THE PRICE DEVIATION IS DIRECTLY PROPOR-TIONAL TO THE SQUARE ROOT OF TIME.
> Hence the investor who wants to sell after the deviation doubles, that is with a difference twice as large between the buy and sell price, must wait four times longer ...
> How astonishing and admirable are the ways of Providence, what thoughts come to our mind when observing the marvellous order which presides over the most minute details of the most hidden events! What! The changes in stock market prices are subject to fixed mathematical laws! Events produced by the passing fancy of men, the most unpredictable shocks of the political world, of clever financial schemes, the outcome of a vast number of unrelated events, all this combines and randomness becomes a word without meaning! And now worldly princes, learn and be humble, you who in your pride, dream to hold in your hands the destiny of nations, kings of finance who have at your disposal the wealth and credit of governments, you are but frail and docile instruments in the hands of the One who brings all causes and effects together in harmony and who, as the Bible says, has measured, weighed and parcelled out everything in perfect order.
> Man bustles but God leads.

51 *"In this model, he started by looking..."*
In Bachelier's words (as translated in Cootner 1964):

At a given instant, the market believes in neither a rise nor a fall of true prices. But if the market believes in neither a rise nor a fall of true prices, it may suppose more or less likely some fluctuations of a given amplitude. The determination of the law of probability consistent with the market at a given instant will be the purpose of this study . . .

The 'mathematical expectation' of an uncertain gain is the product of that gain and the corresponding probability of its occurring. The 'total mathematical expectation' of a player will be the sum of the products of the uncertain gains and the corresponding probabilities of their occurring. Obviously a player will have neither advantage nor disadvantage if his total mathematical expectation is zero. Then the game is called a 'fair game...'

The spot buyer [on an exchange] may be compared with a gambler. In effect, if the price of a security might increase after its purchase, a decrease is equally possible.

52   *"In effect, prices follow…"*

An early reference to the random-walk concept appeared in 1905, in the letters pages of *Nature*, a British scientific journal. Under the headline, "The Problem of the Random Walk," Karl Pearson, a professor and Fellow of the Royal Society, wrote to ask whether any readers could tell him "a solution of the following problem":

A man starts from a point O and walks l yards in a straight line; he then turns through any angle whatever and walks another l yards in a second straight line. He repeats this process n times. I require the probability that after n stretches he is at a distance between r and δr from his starting point, O.

A distinguished scientist, Lord Rayleigh, responded, to which Pearson rejoined:

The lesson of Lord Rayleigh's solution is that in an open country the most probable place to find a drunken man who is at all capable of keeping on his feet is somewhere near his starting point!

Readers with an antiquarian turn of mind can find the correspondence starting on the letters pages of *Nature* 72 (27 July, 1905): 1865. It continues in the following two issues.

Well beyond those works of Pearson and Rayleigh, the year 1905 remains marked by three papers by Albert Einstein, one of which concerns Brownian motion in statistical physics. Random walk and Brownian motion instantly became a core topic of science.

In due time, random walk in the plane found a long-forgotten precursor in John Venn (1834-1923), he of the Venn diagrams of logic. The story is told and illustrated in Chapter H3 (pages 205-207) of Mandelbrot 2002.

54   *"But it was not until 1956..."*
Cowles 1933 was published in his journal, *Econometrica*. A follow-up study eleven years later (Cowles 1944) found his forecasters had not improved with age or experience. Kendall 1953 appeared in the *Journal of the Royal Statistical Society*. And Samuelson's doctoral student mentioned in the text was Richard J. Kruizenga; his thesis was titeld, "Put and call options: A theoretical and market analysis."

55   *"In fact, in 1976 some economists..."*
Rozeff and Kinney 1976.

Chapter IV
The House of Modern Finance

60   *"Clearly, a double-edged sword..."*
The U.S. CAPM study was Graham and Harvey 2001. The European study was Bancel and Mittoo 2003. The regulatory arguments refer to New York State Consumer Protection Board 2001 and Monopolies and Mergers Commission 1997.

61   *"I was never aware of the Great Depression..."*
Recounted on Markowitz's receiving, with Sharpe and Miller, the 1990 Bank of Sweden Prize in Economic Sciences in Memory of Alfred Nobel. The entire account of his work given here is based on his own writings: Markowitz 1959, his Nobel Prize autobiography available at the Nobel e-Museum (http://www.nobel.se/economics/laureates), and his 1999 retrospective article in *Financial Analysts Journal*. Bernstein 1992 is an entertaining account of Markowitz and some of the other founders of financial economics.

63 *"And his ideas spread..."*

In truth, as Markowitz points out, some others had suggested using variance to describe risk before. A. D. Roy, an economist at Cambridge
University, happened to be working on models similar to Markowitz's at
the same time. But Markowitz beat Roy to the publishers and, in subsequent years, fully elaborated his ideas into a practical theory.

65 *"So if you buy a bit of each..."*

The essence of Markowitz's portfolio theory, the so-called mean-vari-
ance criterion, is simple enough: Given two portfolios of investments to
choose from, go for the one with the highest (expected) mean return and
the lowest variance, or risk. But the complication arises in the obvious
practical question: How do you calculate the portfolios' mean and variance?

Calculating the mean is easy: Simply take the expected return for
each stock in a portfolio, and multiply it by its weighting in that portfolio. Thus, for a two-stock portfolio, if you put 40 percent of your money
in stock A, which you expect to return 5 percent, and 60 percent in stock
B, with an expected return of 10 percent, you can expect the portfolio
overall to yield 8 percent (0.4 x 5 plus 0.6 x 10).

But the risk of the stocks, as measured by their variance, does not add
so simply; it can be greater or less than the simple weighted average,
depending on how closely the stocks track each other—their correlation.
Two stocks that tend to crash at the same time are going to make a
riskier portfolio than two stocks that move in opposite directions.
Herewith the formula for the variance of a two-stock portfolio P, where
σ_A and σ_B are the standard deviations of stock A and B, the square being
the variances, w is each stock's weighting in the portfolio, and ρ_{AB} or rho
is the correlation between A and B:

$$\sigma_P^2 = w_A^2 \sigma_A^2 + w_B^2 \sigma_B^2 + 2 w_A w_B \sigma_A \sigma_B \rho_{AB}$$

To see how that works, assume that the volatility, or standard deviation is 10 percent for stock A, 15 percent for stock B. Plugging in the
numbers, the equation simplifies to:

$$\sigma_P^2 = 0.4^2 \cdot 0.10^2 + 0.6^2 \cdot 0.15^2 + 2 \cdot 0.4 \cdot 0.6 \cdot 0.10 \cdot 0.15 \cdot \rho_{AB}$$

or $97 + 72 \rho \cdot 10^{-4}$. Clearly, the higher the correlation, the bigger the variance and risk. So compare portfolios: Say one has stocks A and B that
move in lock-step, with a correlation of 1. Another has stocks A and B

that move opposite to each other, with a correlation of –1. Plugging in the numbers, in the first case the variance is 169 • 10⁻⁴ and the standard deviation (square root of the variance) is 13 • 10⁻²; so you can say the volatility is 13 percent. The other has a variance of 25, and a volatility of 5 percent. Clearly, the more efficient portfolio is the second. Both have an expected return of 8 percent, but the second carries far less risk than the first—that is, it is more likely to deliver the profits.

As you start adding securities to the portfolio, the calculations lengthen—but the principle remains the same. You can then plot all the possible portfolios on graph paper and see which offer the optimal return for the minimal risk. Of course, all the calculations assume the bell-curve math of mean and variance is relevant to markets.

For readers seeking more information, there are many elementary textbooks on investment management and theory. A good one, requiring little mathematical background, is Bodie, Kane and Marcus 2002. For those with more math, Watsham and Parramore 1997 provides a broad overview of this and many other aspects of mathematical finance.

66   *"The answer to the number-crunching..."*
Recounted on Sharpe's receiving, with Markowitz and Miller, the 1990 Nobel for economics. The account of Sharpe's work presented here comes from several primary sources, including his Nobel autobiography and address, Sharpe 1964 paper, recollections in Markowitz 1999, and the transcript of a 1998 interview with Sharpe in *Dow Jones Asset Manager*.

68   *"That is a lot to expect..."*
The "expected return-beta" equation is the heart of Sharpe's model. As described in the text, it reads:

$$E(r_i) = r_f + \beta_i(E(r_M) - r_f).$$

E stands for the mathematical expectation operator. It means a projected outcome times the odds of that outcome happening. For instance, the expectation of a fair coin-tossing game is 0, because you have a 0.5 chance of winning 1 point and a 0.5 chance of losing 1 point (0.5 times 1 plus 0.5 times –1 equals 0). So the present equation is saying that the expected return *r* on security *i* equals the sum of two numbers. The first is the "risk-free rate" that you would expect to get from something safe like a Treasury bill. The second is Sharpe's beta times the "market premium"—that is, however much better you expect the market *M* to perform over the Treasury rate. Beta is the key to it. Each stock has its own beta, or degree to which its price movements correlate to that of the market overall. It is defined as how much the stock varies with the market— the covariance—divided by the variance or risk of the market itself. Again, this is all bell-curve math: Its validity depends entirely on whether prices really do fit the bell curve.

70   *"But today most economists credit Sharpe…"*
Jack Treynor of Arthur D. Little Inc. also had circulated research on the subject. But he gets scant credit because he did not publish—in part, according to a former ADL colleague, Fischer Black, because it "never quite satisfied the perfectionist in him, and in part (I believe) because he did not have an academic job." See Black 1989.

71   *"As fate would have it…"*
The source for this is the contemporaneous, if sparse, reporting of the *Wall Street Journal*. In hindsight, the newspaper appears to have under-estimated, and thus under-played, the importance of the exchange's opening.

72   *"The answer came…"*
From the eulogy of Black, Scholes 1995. The account of their discovery given here is based on the published recollections of the participants, including Black 1989, Scholes 2001, and the autobiographical essays Merton and Scholes 1997.

73   *"The Black-Scholes formula permitted…"*
The Black-Scholes formula looks complex, but working with it is a simple matter of plugging numbers into their proper places in a spreadsheet or calculator. The price of a call option to buy a stock at a specific price and time is:

$$C_0 = S_0 N(d_1) - Xe^{-rT} N(d_2)$$

Here, C_0 is the price of the call option; S_0 is the current stock price; X is exercise price at which the option contract allows you to buy the stock; r is the risk-free interest rate; and T is the time to maturity. The other two functions, $N(d_1)$ and $N(d_2)$, are the probabilities of a random number, d, that follows a bell-curve distribution, being less than the quantities below:

$$d_1 = \frac{\ln(S_0 / X) + (r + \sigma^2 / 2)T}{\sigma\sqrt{T}}$$

$$d_2 = d_1 - \sigma\sqrt{T}$$

where σ is the standard deviation of the stock price and ln is the natural logarithm. They are, in essence, the probabilities of the option expiring "in the money," that is, paying off.

As illustration, we can borrow an example from a popular textbook, Bodie, Kane and Marcus 2002: Assume the current stock price S_0 is $100, the exercise price X is $95, the risk-free rate is 10 percent, the time to expiration T is a quarter-year, and the stock's standard deviation is 50 percent. A calculator quickly shows d_1 is 0.43 and d_2 is 0.18. A bell-curve probability table shows $N(d_1)$ is 0.6664 and $N(d_2)$ is 0.5714. Finally, plugging those values into the full equation, we find the fair price of the call option C_0 is $13.70. Again, this all assumes the bell-curve math applies to stock prices.

Chapter V
The Case Against the Modern Theory of Finance

82   *"The result? If you had followed ... "*
Reported in James 2003, 2004.

85   *"In computer simulations... "*
De Grauwe and Grimaldi 2003.

91   *"Logarithms rescale everything"*
Here are a few comments for readers who want a quick reminder of logarithms. The decimal logarithm, a function now automated by a button on many pocket calculators, takes the number you input and writes it in a different form. Roughly speaking, it is an order of magnitude: the part of the logarithm before its decimal point is the number of digits in the input, minus one. When the input lies between 0 and 1, this rule generalizes to yield a negative output. Specifically, the logarithm gives the power by which 10—or another number above 1 called the base—would have to be raised to get back to the input number.

For instance, 100 is 10 squared – so the base, 10, would have to be raised by a power of 2 to get the inputted number, 100, back again. Or, in the standard notation, log 100 = 2. Likewise, 1,000 is 10 to the third power, so log 1,000 = 3. An intermediate case: 400 is partway between 100 and 1,000, so it is 10 to some power between 2 and 3—or, to be precise, 2.6; so log 400 = 2.602. One more step: The most convenient logarithm base is not 10 but a number called e, which—as mentioned earlier—enters in the formula for the Gaussian distribution. It begins by 2.71828; its digits continue indefinitely, without ever repeating. It is a peculiar number, but happens to be quite important throughout mathematics and also in finance. It affects, for instance, how interest compounds continuously on a bond, savings account or mortgage.

Why bother with logarithms, to any base? Because expressing a number in logarithms rescales it so that, rather than focusing on the size of the number as we normally do, we can more easily compare it to other numbers nearby. Thus, $1 price jumps from $10 to $11and from $1,000 to $1,001 are equal on the dollars scale but the logarithmic scale shows the former to be more important than the latter.

95   *"When Cootner of MIT..."*
Cootner 1964.

96   *"My student, Eugene Fama..."*
Fama 1964, revised and published as Fama 1965b.

96   *"They call it kurtosis..."*
Kurtosis is one of the founders of the standard measures of a distribution curve's shape, which are based on the first four "moments." The first moment is the average value; the second is the variance; third is the skewness—a measure of how asymmetrically the data are distributed around the average; and fourth is kurtosis, a measure of how tall or squat the curve is. A bell curve has a kurtosis of three. Larger values imply the curve is tall in the center, with fat tails.

98   *"The same phenomenon..."*
See, for instance, an analysis of mean, variance, skewness, and kurtosis in Deutschmark, yen, pound, French franc, and Swiss franc currency crosses against the dollar, from 1987 to 1996, at p. 73 of Adler, Feldman and Taqqu 1998. The Citibank study referred to is James 2002.

98   *"Several other studies have found…"*
For instance, Lo and MacKinley 1988.

101 *"Anomaly 1:* The P/E Effect…"
For instance, Basu 1983.

101 *"Anomaly 2:* The Small-Firm-in-January Effect…"
Banz and Breen 1986.

104 *"Those tools, part of a…"*
GARCH stands for Generalized Auto-Regressive Conditional Heteroskedasticity, a mouthful in my language. It refers to a set of statistical tools
to model data whose variability changes with time ("heteroskedastic" in
statistics terminology). The "auto-regressive conditional" term means the
changes in variability are controlled by the data's own past behaviour. And
"generalized" means the model has been broadened to accommodate more
circumstances than when initially developed in 1982 as ARCH.

107 *"Our whole focus is on the extremes…"*
Meriwether was quoted in "Long Term Capital Chief Acknowledges
Flawed Tactics," by Gregory Zuckerman, in the *Wall Street Journal*, August
21, 2000. Scholes 2000 describes its author's views of the LTCM debacle.

Chapter VI
Turbulent Markets: A Preview

112 *"Here, in an illustration…"*
That paper was Mandelbrot 1972.

114 *"From the* Notebooks…"*
Quoted in Masters 1999.

116 *"I coined its name…"*
A word was needed without delay because a first book on the field had
been completed (it was to become Mandelbrot 1975) and the publisher
insisted on its having a title.

117 *"To avoid misunderstanding…"*
These cartoons were sketched in Mandelbrot 1997a and developed in
Mandelbrot 2001c.

Chapter VII

Studies in Roughness: A Fractal Primer

138 *"The curve is crinkly..."*

Fractal dimension is an intricate topic, as you might expect. There are several variant definitions, each suited to a different purpose. For tidy fractal patterns like the Koch curve and Sierpinski gasket that scale uniformly in all directions than dimensions layerly coincide, the simplest is the "similarity dimension" whose formula is:

$$d_s = \log(N)/\log(1/r)$$

where r is the ratio by which the measuring unit scales up or down, and N is the number of measuring units needed to finish the job. In the Koch case, we said $r = 3$ and $N = 4$, so the similarity dimension is $\log 4 / \log 1/3$ = 1.2618....

A more versatile formula is for the "box-counting" dimension. As the name implies, you get it by counting how many boxes, or squares, of different sizes are required to enclose a fractal pattern. Looking at the Koch curve again, you can see that, if you start out trying to cover the curve with boxes one-third the width of the object, three boxes are needed. (In the text below, $N(r_1)$ stands for the number of boxes of radius r_1 required.) Shrink the boxes by one-third again, to one-ninth the original shape, and twelve are needed. Shrink again, and forty-eight are needed.

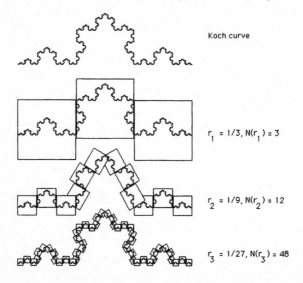

Koch curve

$r_1 = 1/3, N(r_1) = 3$

$r_2 = 1/9, N(r_2) = 12$

$r_3 = 1/27, N(r_3) = 48$

A pattern soon emerges, and it fits this formula:

$$d_b = \lim_{r_n \downarrow 0} \frac{Log N(r_n)}{Log(1/r_n)}$$

where *lim* stands for the mathematical limit approached by the ratio of logarithms as the reduction ratio r gets closer to 0. The answer, as with the similarity dimension, is log 4/log 3 = 1.2618. But the box-counting dimension can be used on a wider range of fractal objects. There are many other types of dimension—mass, Hausdorff, packing. A good primer on this can be found at the Yale Math 190 Web site mentioned earlier (from which the above illustration is reprinted).

There are more-complicated ways of calculating fractal dimension, and many different ways of expressing it. But scientists have found it a useful tool for measuring all manner of phenomena—from the roughness of a metal fracture to the variability of a financial chart.

146 *"Its perplexing mix of simplicity and complexity..."*
The Mandelbrot set starts with an old problem that by Gaston Julia, my professor of differential geometry at the École Polytechnique in Paris has studied in his youth. As fate would have it, my mathematician uncle tried to get me to take up this topic for a doctoral thesis. I had other plans at the time, but eventually returned to the topic. The problem concerns so-called iterated functions, a kind of mathematical feedback loop that keeps operating on its own output again and again. For instance, in the function

$$z_1 = z_0^2 + c$$

z_0 is the starting value of the process, c is a constant, and z_1 is the first output. Then repeat the operation:

$$z_2 = z_1^2 + c$$

and

$$z_3 = z_2^2 + c$$

If you keep doing this forever with starting numbers c like 3 or 4, the sequence (irrespective of z_0) will soar off into infinity. But if you say z

and c are "complex" numbers, with the imaginary term $\sqrt{-1}$ in them, then the story gets more interesting. (The name, imaginary, harks back to times when those numbers were not understood, but these are very real-world numbers used in many realms of science and engineering.) Sometimes the series will veer off towards infinity—but sometimes it will not. And the precise pattern is exquisitely intricate.

With the Mandelbrot set, you start by setting z_0 equal to 0, and then see what happens to the sequence when you try different values of c. If the sequence does not run away to infinity, then c is said to be in the Mandelbrot set. If it does, c is not in the set. Black and white illustrations of the set typically assign a computer-screen pixel to every value of c being tested, and then paint it black if the pixel is inside the Mandelbrot set and a variety of other colors if it is not. Different colors are often used to denote how quickly the series soars to infinity. The surprising thing is that as you look at smaller and smaller scales—say, zoom in on values of c in a tenth of the screen rather than the whole screen—you find the pattern of what is in the set and what is not becomes far more complicated than it at first appeared. Zoom again, and yet more fine detail emerges. You can do this forever, and at each stage get an entirely different picture. Its study has become a classic problem in pure mathematics.

The Mandelbrot set belongs to both fractal geometry and chaos theory. A chaotic system, far from being disorganized or non-organized, starts with one particular point and cranks it through a repeating process; the outcome is unpredictable if you do not know the process—and it depends heavily on the starting point. The most famous example of chaos was proposed by meteorologist Edward Lorenz in 1972: Can the flap of a butterfly's wings in Brazil set off a tornado in Texas? The basic idea is that if you stand a pencil on its point and let it fall through force of gravity, exactly where it lands depends on where it began, whether it was leaning infinitesimally in one direction or another.

Chapter VIII
The Mystery of Cotton

147 *"Central mysteries of finance"*
The source book on long tails is Mandelbrot 1997a.

152 *"Zipf, eyeballing his charts..."*
Zipf ranked words by their frequency. The most common word in a text

gets a rank of 1, the second-most common a rank of 2, and so on. Then a formula gives the probability of each word occurring in a text:

$$Q(r) \approx Fr^{-1/\alpha}$$

where Q is the probability distribution function, r is the ranking, F is a constant that Zipf estimated at one-tenth, and $1/\alpha$ is the critical power-law factor. The bigger the value of α, the richer the vocabulary—that is, the curve plotting the frequency of each word against its ranking declines more gently, so rare words happen more often than they otherwise might. Zipf asserted that α is 1. In fact, there are many empirical problems with Zipf's "law," as some call it; not least of them is that it simply does not accurately reflect what happens in real language. But it proved interesting mathematically—and led me to a generalization called Zipf-Mandelbrot law, and then—step by step—to many other, more fruitful studies in power laws.

153 *"One of Pareto's equations…"*
From Pareto 1905. His books, even today, stand as models of clear thinking and powerful writing.

155 *"To me—I did not even study…"*
Pareto's formula, in algebraic notation, is:

$$P(u) = (u/m)^{-\alpha}$$

In English: What proportion, P, of people earn more than some level of income, u? The answer is on the right side. To illustrate it, take the specific example in the text: What percentage of people in the workforce earn at least ten times the U.S. minimum income = $10,712 a year? Divide u, which in this case is $107,120, by the minimum income, m. So the ratio in brackets is easy to calculate: u/m is 10. Then the formula says to raise that ratio to a special power—the "power" in this power law—called minus alpha, or $-\alpha$; Pareto estimated $-\alpha$ as -3/2. Raising something to a negative power of 3/2 means that you first cube it, then take the square root; and then you invert it all—that is, divide it into 1. Here, 10 cubed is 1,000. Its square root is 31.6. One divided by 31.6 is 0.032, or 3.2 percent. So according to Pareto's formula, the answer should be: 3.2 percent of everyone making more than the minimum wage is taking home $107,120 a year.

158 *"Insurance claims make a particularly good..."*
    For instance, Bencker and Sternberg 1957.

161 *"The key parameter is alpha..."*
    As you might expect from a something that covers so broad a range of
    behavior, the L-stable formula is a complicated affair. For the curious, its
    characteristic function is:

$$\log f(t) = i\delta t - \gamma \,|\, t \,|^{\alpha} \,[1 + i\beta(t/|t|)\tan(\alpha\pi/2)]$$

    It means that the L-stable probability distributions have four parame-
    ters—the key variables that decide what the final shape of the curve will
    be. These are the four "knobs" that, by tuning, determine whether we
    are describing a bell curve, a Pareto-style curve, or something entirely dif-
    ferent. The "location" parameter is δ. The "scale" parameter is γ, mean-
    ing it determines the magnitude of the probabilities overall. The index of
    skewness is β: If it is 0, the curve is symmetrical. And the most important
    parameter is α, which determines the fatness of the tails. When α is 2 and
    β is 0, the equation describes the standard bell curve. When α is 1 and β is
    0, we have the Cauchy distribution with its very fat tails.

161 *"So, many seemingly unrelated..."*
    Those papers relative to the distribution of income are listed in the bibli-
    ography. They appeared in 1959, 1960, 1961, and 1962a, and are repro-
    duced in Mandelbrot 1997a.

Chapter IX
Long Memory, from the Nile to
the Marketplace

173 *"financial markets work..."*
    The source book on long dependence is Mandelbrot 2002.

174 *"About why the Nile behaves..."*
    From Book II of Herodotus's *The Histories*.

174 *"There has been a great deal..."*
    From Cole 1980. Also, all of the hydrological data on the Nile cited here
    are either from Hurst's own writings, or from Shahin 1985.

178 *"A strange number; but it was…"*

Hurst's formula started by calculating the average annual rainfall or water discharge from a river, and keeping a running tally, year by year, of the accumulated deviations from that average. So if, as in the New York case, the average discharge is forty-two inches a year, but in the first year the actual discharge is forty-three, in the second year forty-five, in the third year thirty-six and in the fourth year forty-four, then the accumulated deviations each year are one, then four, then minus two, then zero. He then looked at the peak value that that accumulation reached (four, in this example) and compared it to the lowest level it reached (here, minus two.) He called the difference, six, the range, or R. His formula gives the value of R, which indicates how big the reservoir should be to avoid floods or droughts downriver. It is determined by σ, the standard deviation of the discharges from one year to the next; N, the number of years under study; and a, the power-law exponent that drives the whole equation. Hurst, using logs, used the equation:

$$\log(\frac{R}{\sigma}) = K \log(\frac{N}{2})$$

or, without the logs:

$$R = \sigma(\frac{N}{2})^K$$

Based on his research, he estimated K to be 0.73, with a standard deviation of 0.09 and ranging between 0.46 and 0.96 (Hurst 1951). So, plugging in the numbers for the New York case: With a standard deviation of 6.28, K of 0.72, and an N of 100, R equals the 105 inches cited in the text. I found it necessary to correct Hurst's formulation and introduced a more appropriate exponent I called H. It does not affect the main point, namely that the R for river discharges and other natural phenomena that Hurst studied grows more quickly as time passes than it would if he were studying a simple random process like a coin-toss game. In his words:

> Although many natural phenomena have a nearly normal frequency distribution this is only the case when their order of occurrence is ignored. When records of natural phenomena extend over long periods there are considerable variations both of means and

standard deviations from one period to another. The tendency to occur in groups makes both the mean and the standard deviation computed from short period of years more variable than is the case in random distributions.

191 *"More common is the risk-avoiding bureaucrat..."*
In fact, Langbein went on to publish not one, but four papers by Mandelbrot and Wallis in 1968 and 1969.

192 *"But shortly after, other economists said..."*
The first published rebuttal was Willinger, Taqqu, and Teverovsky 1999.

Chapter X
Noah, Joseph, and Market Bubbles

202 *"To separate the two effects, I developed..."*
The R/S, or rescaled-range statistic, is widely used now for testing whether long-term dependence is present in a series of data. One of its principal virtues is that, in contrast to many common statistical tests, it makes no assumptions about how the original data are organized—a critical point when studying something like stock prices for which evidence abounds that the conventional assumptions are flatly wrong. The cookbook R/S formula simply measures whether, over varying periods of time, the amount by which the data vary from maximum to minimum is greater or smaller than what you would expect if each data point were independent of the last. If different from expectations, then the precise sequence of the data must be important: A "run" of gains or losses must be pushing the extreme values farther than they would otherwise go by pure chance. The equation for calculating it:

$$\frac{Max_{1\le k\le n}\sum_{j=1}^{k}(r_j - \bar{r}_n) - Min_{1\le k\le n}\sum_{j=1}^{k}(r_j - \bar{r}_n)}{\left[\frac{1}{n}\sum_j (r_j - \bar{r}_n)^2\right]^{1/2}}$$

To explain: Start by looking at the return r—the profit or loss from, say, a stock-price movement—over different time periods of a day, two days, three days, and so on up to the full time-series of, say, one hundred days or n. Calculate the average return, r_n, over the entire one hundred days. Then for each shorter time-period—a day, two days, and so on—calculate the difference between the return r_j over that period and the average return, r_n, over one hundred days, and keep a running total of all the differences as the time-periods lengthen up to a period k. Do this for just one day ($k = 1$); then two days ($k = 2$); and so on until $k = 100$. Then take the maximum, or *Max*, of all those differences. Likewise, find the minimum, or *Min*, of all the differences. Subtract one from the other, to get an estimate of the range from peak to trough in the accumulated deviations. That is the numerator. The denominator is a conventional measure of the standard deviation in the data series.

If the data were independent, you would expect the numerator and denominator to be in a ratio of 1:2—or $H = 1/2$. Any value other than that implies the presence of long-term dependence. If the range is bigger than expected, and $H > 1/2$, then the data are "persistent" and there are long runs. If the range is smaller and $H < 1/2$, then the data are "anti-persistent" and the values have a tendency to keep doubling back on themselves.

Further elaboration can be found in the original Mandelbrot-Wallis papers listed in this book's bibliography, in this book's Web site, and in Peters 1996.

Chapter XI
The Multifractal Nature of Trading Time

209 *"power-law distribution"*

This key observation was a total surprise and greatly impressed me. It led me, in a comment concluding Mandelbrot 1972, to remark that the techniques being developed for turbulence would also apply in economics.

209 *"as early as 1975"*

The first multifractal models of price variation were the cartoons to be discussed momentarily and the fractional Brownian motions in multifractal trading time to be discussed starting on page 127. They are closely related and were first presented in Mandelbrot 1997a; see also Mandelbrot 1991a b c d e. The first tests were reported in Mandelbrot, Calvet, and Fisher 1997, Calvet, Fisher, and Mandelbrot 1997, and Fisher, Calvet, and Mandelbrot 1997.

216 *"In fact, this concept…"*

See Mandelbrot and Taylor 1967.

216 *"I co-authored in 1967"*

In that paper, trading time was not taken to be multifractal, but fractal—but neither term was used because they had not yet been coined. That is, the best I could do in 1967 was to consider the increments of trading time as statistically independent, hence to model the Noah but not the Joseph Effect. The novelty I reported in 1972 was that the Noah and Joseph Effects could be united in an intrinsic fashion.

217 *"market behavior"*

Originally, the function $f(\alpha)$ arose in Mandelbrot 1972, 1974, as the logarithm (suitably scaled) of a basic probability. Later on, $f(\alpha)$ was called a "spectrum" of dimension or of singularity. In many cases $f<0$ for some α. Such "negative dimensions" turn out to be indisputable as providing a measure of "degree of emptiness," Mandelbrot 2003.

217 *"The trading time process is expressed…"*

The binomial time bending illustrated on 215 is very much oversimplified. Early on, Mandelbrot 1974ab described much more general cascades. Among further explicit examples of multifractal bending, several are recent ones that I co-authored with Julien Barral. They are available on my web site www.math.yale.edu/mandelbrot.

222 *"And, as you try to work with the model…"*

Both GARCH and multifractal model include a multitude of parameters. In favor of GARCH: it is a combination of concepts long familiar to statisticians. Against GARCH: it denies the existence of long dependence except if it is added to earlier ingredients to form a hybrid called FIGARCH. Also, the parameters estimated from weekly and daily data, when used to create artificial samples, yield time series of completely different character. In favor of multifractals: their parsimony, the fact that the "turbulent" behavior is not deliberately inputted but obtained as output of simplet interpolative cartoons. Multifractals should not be viewed as an "ad-hoc" structure but as the natural counterpart of two classical tools; the generating function (that is, the sequence of moments) and spectral analysis. Their parameters are intrinsic.

Chapter XII
Ten Heresies of Finance

230 *"Consider the so-called Equity Premium Puzzle..."*
A good summary of their initial paper, and the difficulty it had in getting published, is provided in Mehra and Prescott 2003.

231 *"The same reasoning..."*
For more on this, see Babeau, André and Sbano 2002.
    In fact, the precise asset allocation recommendations can vary from that 25-30-45 mix, depending on what the market is doing at any particular time.

232 *"The ultimate fear..."*
See Embrechts, Klüppelberg and Mikosch 1997.

234 *"Concentration is common..."*
See Lantsman, Major and Mangano 2002.

235 *"One day when I was working..."*
Alexander's "Filter" method attracted a great deal of attention–and similar methods have been devised and tried since his day. Unlike many, however, Alexander had the grace to retract when he was wrong. His initial paper was Alexander 1961. The retraction was Alexander 1964. A good discussion of the affair was presented in: Fama and Blume 1966.

238 *"The SEC later sued..."*
Reported in Damato 2004.

248 *"Of course, you cannot predict..."*
See Maillet and Michel 2003.

249 *"The next step is forecasting..."*
See Zumbach, Olsen and Olsen 2000.

Chapter XIII
In the Lab

257 *"A more accurate metaphor..."*
More information on Olsen's views of finance and his funds is available

at his Web sites, http://www.oanda.com and http://www.olsen.ch. A more formal presentation of his fractal views is in Dacarogna *et al.* 2001. A brief summary is: Müller *et al.* 1993.

260 *"He and some colleagues..."*
The paper on "tail chiseling" is Bouchaud *et al.* 1998. More information on the funds is at http://www.science-finance.fr.

261 *"But the research has moved on..."*
Peters' investment firm is at http://www.panagora.com. His most recent fractal book is Peters 1996.

262 *"For instance, one study..."*
See Richards 2000.

264 *"Others have used this technique..."*
More information on this kind of "driven iterated function system" approach is at the Yale Math 190 Web site, http://classes.yale.edu/fractals/index.html

266 *"In 2000, some researchers in ..."*
See Belkacem, Véhel, and Walter 1999. See also Fama 1965b.

268 *"In the foreign exchange market..."*
See Batten and Ellis 1999.

270 *"This method, developed by..."*
Madan, Carr, and Chang 1998.

271 *"By any measure, the late 1990s..."*
See Summers 2000.

276 *"On the night of February 1, 1953..."*
Bassi, Embrechts, and Kafetzaki 1998.

Bibliography

Acar, Emmanuel and Andrew Pearson. 2001. Distribution of returns generated by stochastic exposure; an application to VaR calculation in the futures markets. *AFIR Colloquium*, Toronto Sept. 6–7. International Actuarial Association.

Adelman, Irma. 1965. Long cycles—fact or artifact? *American Economic Review* 55 (3): 444–463.

Adler, Robert J., Raise E. Feldman and Murad S. Taqqu, *eds*. 1998. *A Practical Guide to Heavy Tails: Statistical Techniques and Applications*. Basel: Birkhäuser.

Alexander, Gordon J. and William F. Sharpe. 1989. *Fundamentals of Investments*. Englewood Cliffs, NJ: Prentice-Hall International Inc.

Alexander, Sidney S. 1961. Price movements in speculative markets: Trends or random walks? *Industrial Management Review* 2 (2): 7-26.

Alexander, Sidney S. 1964. Price movements in speculative markets: Trends or random walks, number 2. *Industrial Management Review* 5 (2): 25-46.

Alligood, Kathleen T., Tim D. Sauer, and James A. Yorke. 1996. *Chaos: An Introduction to Dynamical Systems*. New York: Springer-Verlag.

Alvarez-Ramirez, Jose, Myrian Cisneros, Carlos Ibarra-Valdez, and Angel Soriano. 2002. Multifractal Hurst analysis of crude oil prices. *Physica A* 313: 651-670.

Babeau, André and Teresa Sbano. 2002. *Household Wealth in the National Accounts of Europe, the United States and Japan*. Paris: Organization for Economic Cooperation and Development.

Bachelier, Louis. 1900. Théorie de la Spéculation. Doctoral dissertation.

Annales Scientifiques de l'École Normale Supérieure (iii) 17, 21-86. Translation: Cootner, 1964.

Bakshi, Gurdip and Dilip Madan. 1998. What is the probability of a stock market crash? *NYU Conference on Finance and Accounting.*

Baillie, Richard T. 1996. Long memory processes and fractional integration in econometrics. *Journal of Econometrics* 73: 5-59.

Baillie, Richard T. and Maxwell L. King. 1996. Editors' introduction: Fractional differencing and long memory processes. *Journal of Econometrics* 73: 1-3.

Bancel, Franck and Usha R. Mittoo. 2003. The determinants of capital structure choice: A survey of European firms. Conference report, American Finance Association, January.

Banz, R.W. and W. Breen. 1986. Sample dependent results using accounting and market data: Some evidence. *Journal of Finance* 41: 779-793.

Basel Committee on Banking Supervision. 2003. *Consultative document: Overview of the New Basel Capital Accord.* At www.bis.org.

Bassi, Franco, Paul Embrechts, and Maria Kafetzaki. 1998. Risk management and quantile estimation. Adler et al. 1998.

Basu, S. 1983. The relationship between earnings' yield, market value and return for NYSE common stocks: Further evidence. *Journal of Financial Economics* 12: 129-156.

Batten, Jonathan and Craig Ellis. 1999. Volatility scaling in foreign exchange markets. *CREFS Centre Working Papers* 99-04. Singapore: Nanyang Technological University.

Belkacem, Lotfi, Jacques Lévy Véhel and Christian Walter. 1999. CAPM, risk and portfolio selection in "?-stable markets." *Fractals* 8 (1): 99-115.

Bell, E.T. 1937. *Men of Mathematics.* New York: Simon & Schuster Inc.

Bencker, L.-G. and I. Sternberg. 1957. An attempt to find an expression for the distribution of fire damage amount. *Transactions 15th International Congress of Actuaries* 2: 288-294.

Bernstein, Peter L. 1992. *Capital Ideas: The Improbable Origins of Modern Wall Street.* New York: Free Press.

Black, Fischer. 1989. How we came up with the option formula. *Journal of Portfolio Management* Winter: 4-8.

Black, Fischer and Myron Scholes. 1973. The pricing of options and corporate liabilities. *Journal of Political Economy* 81 (May/June): 637-654.

Bodie, Zvi, Alex Kane, and Alan J. Marcus. 2002. *Investments.* 5TH Edition. New York: McGraw-Hill/Irwin.

Bollerslev, Tim. 1987. A conditionally heteroskedastic time series model for speculative prices and rates of return. *Review of Economics and Statistics*: 542-547.

Bouchaud, J.P. D. Sornette, C. Walter, and J.P. Aguilar. 1998. Taming large events: Optimal portfolio theory for strongly fluctuating assets. *International Journal of Theoretical and Applied Finance* 1 (1): 25-41.

Bouchaud, Jean-Philippe and Marc Potters. 2000. *Theory of Financial Risks: From Statistical Physics to Risk Management.* Cambridge, U.K.: Cambridge University Press.

Bouchaud, Jean-Philippe and Marc Potters. 2001. Welcome to a non-Black-Scholes world. *Quantitative Finance* 1 (5): 482-483.

Bouchaud, Jean-Philippe. 2002. An introduction to statistical finance. *Physica A* 313: 238-251.

Buffett, Warren E. 1988. To the Shareholders of Berkshire Hathaway Inc. *Annual Report.* Omaha, Neb.: Berkshire Hathaway Inc.

Burton, Jonathan. 1998. Revisiting the capital asset pricing model. *Dow Jones Asset Manager* May-June: 20-28.

Calvet, Laurent, Adlai Fisher, and Benoit Mandelbrot. 1997. Large deviations and the distribution of price changes. *Cowles Foundation Discussion Paper* 1165 (September).

Calvet, Laurent and Adlai Fisher. 2002. Multifractality in asset returns: Theory and evidence. *Review of Economics and Statistics* 84 (3): 381-406.

Campbell, John Y., Andrew W. Low, and A. Craig MacKinlay. 1997. *The Econometrics of Financial Markets.* Princeton, NJ : Princeton University Press.

Cavalcante, Jorge and A. Assaf, 2002. Long range dependence in the returns and volatility of the Brazilian stock market. Conference report, 24th Encontro Brasileiro de Econometria, December 11-13.

Chernoff, Joel. 2002. The house that Harry built: Physics, math and managing money: Modern portfolio theory at 50. *Pensions & Investments* April 29.

Cheung, Yin-Wong and Kon S. Lai. 1995. A search for long memory in international stock market returns. *Journal of International Money and Finance* 14 (4): 597-615.

Cheung, Yin-Wong and Kon S. Lai. 2001. Long memory and nonlinear mean reversion in Japanese yen-based real exchange rates. *Journal of International Money and Finance* 20: 115-132.

Cole, J.A. 1980. H.E.Hurst, D.Sc., CMG (1880–1978). *Journal of Hydrology* 46: 1-3.

Commodity Futures Trading Commission. 2003. *Futures Commission Merchant Reports for 2003.* On the Web at http://www.cftc.gov/tm/tmfcm.htm.

Cootner, Paul H, ed. 1964. *The Random Character of Stock Market Prices.* Cambridge, MA.: MIT Press.

Courtault, Jean-Michel. 2000. Louis Bachelier: On the centenary of Théorie de la Spéculation. *Mathematical Finance* 10 (3) July: 339-353.

Courtault, Jean-Michel et al. 2000. Louis Bachelier: Fondateur de la finance mathématique. A Web site, sponsored by the Université de Franche-Comté, publishing primary manuscripts and photographs of Bachelier's life and times, for the centenary of his doctoral thesis: http://sjepg.univ-fcomte.fr/La_recherche/Libre/bachelier/page01/page01.htm.

Cowles, Alfred. 1933. Can stock market forecasters forecast? *Econometrica* 1, July: 309-324.

Cowles, Alfred. 1944. Stock market forecasting. *Econometrica* 12: 206-214.

Dacarogna, Michel M., Ramazan Gençay, Ulrich A. Müller , Richard B. Olsen, and Olivier V. Pictet. 2001. *An Introduction to High-Frequency Finance*. San Diego, CA: Academic Press.

Damato, Karen. 2004. One Case Shows the Perils of "Smoothing." *Wall Street Journal* March 24.

De Grauwe, P. and M. Grimaldi. 2003. Bubbling and crashing exchanges rates. *CESIfo Working Paper* 1045.

Desai, Padma. 2000. Why did the ruble collapse in August 1998? *American Economic Review* 90 (2): 48-52.

De Vries, Casper G. 2001. Fat tails and the history of the guilder. *Tinbergen Magazine* 4 (Fall): 3-6.

Diebold, Francis X. and James A. Nason, 1990. Nonparametric exchange rate predictions. *Journal of International Economics* 28: 315-332.

Dusak, Katherine. 1973. Futures trading and investor returns: An investigation of commodity market risk premiums. *Journal of Political Economy* 81: 1387-1406.

Eglash, Ron. 1999. *African Fractals: Modern Computing and Indigenous Design*. New Brunswick: Rutgers University Press.

Elie, L., N. El Karoui, T. Jeantheau, and A. Pferzel. 1993. Les modèles ARCH sur les cours de change. *AFIR Colloquium*, Rome, International Actuarial Association.

Embrechts, Paul, Claudia Klüppelberg, and Thomas Mikosch. 1997. *Modelling Extremal Events for Insurance and Finance*. Berlin: Springer-Verlag.

Embrechts, Paul. 2002. Where mathematics, insurance and finance meet. *Quantitative Finance* 2 (6): 402-404.

Engel, Charles and James D. Hamilton. 1990. Long swings in the dollar: Are they in the data and do markets know it? *American Economic Review* 80 (4) September: 689-713.

Falconer, Kenneth. 1990. *Fractal Geometry: Mathematical Foundations and Applications*. Chichester, U.K.: John Wiley & Sons.

Fama, Eugene F. 1964. *The distribution of daily differences of stock prices: a test of Mandelbrot's stable paretian hypothesis.* Ph.D. dissertation, University of Chicago Graduate School of Business.

Fama, Eugene F. 1965a. Portfolio analysis in a stable Paretian market. *Management Science* 2 (3) January: 404-419.

Fama, Eugene F. 1965b. The behavior of stock-market prices. *Journal of Business* 38 (1): 34-105.

Fama, Eugene F. 1991. Efficient capital markets: II. *Journal of Finance* 46 (5): 1575-1617.

Fama, Eugene F. and M. Blume. 1966. Filter rules and stock-market trading. *Journal of Business* 39: 226-241.

Fama, Eugene F. and Kenneth R. French. 1988. Permanent and temporary components of stock prices. *Journal of Political Economy* 96: 246-273.

Fama, Eugene F. and Kenneth R. French. 1992. The cross-section of expected stock returns. *Journal of Finance* 47 (2) June: 427-465.

Feller, W. 1950. *An Introduction to Probability Theory and Its Applications.* New York: Wiley.

Fillol, Jérôme, 2003. Multifractality: theory and evidence. An application to the French stock market. *Economics Bulletin* 3 (31): 1-12.

Financial Executives Research Foundation. 2003. *Valuing Employee Stock Options: A Comparison of Alternative Models.* Research report available at: http://www.ferf.org.

Fisher, Adlai, Laurent Calvet, and Benoit Mandelbrot. 1997. Multifractality of Deutschemark/US dollar exchange rates. *Cowles Foundation Discussion Paper* 1166.

Frame, Michael and Benoit B. Mandelbrot. 2002. *Fractals, Graphics and Mathematics Education.* Washington, D.C.: Mathematical Association of America.

Gleick, James. 1987. *Chaos: Making a New Science.* New York: Viking Penguin.

Gleria, Iram, Raul Matsushita, and Sergio Da Silva. 2002. Scaling power laws in the Sao Paulo Stock Exchange. *Economics Bulletin* 7 (3): 1-12.

Gnedenko, B.V. and A.N. Kolmogorov. 1954. *Limit Distributions for Sums of Independent Random Variables.* English translation by K.L. Chung. Reading, MA: Addison Wesley.

Graham, John R. and Campbell R. Harvey. 2001. The theory and practice of corporate finance: Evidence from the field. *Journal of Financial Economics* 61: 1-52.

Graham, John R. and Campbell R. Harvey. 2002. How do CFOs make capital budgeting and capital structure decisions? *Journal of Applied Corporate Finance* 15 (1): 8-23.

Granger, C.W.J. 1966. The typical spectral shape of an economic variable. *Econometrica* 34 (1): 150-161.

Grimmett, Geoffrey and David Stirzaker. 2002. *Probability and Random Processes*. Third Edition. New York: Oxford University Press.

Hall, Tord. 1970. *Carl Friedrich Gauss: A Biography*. Cambridge, MA: MIT Press.

Harvey, C.R. 1991. The world price of covariance risk. *Journal of Finance* 46: 111-157.

Hermite, C. and T.J. Stieltjes. 1905. *Correspondance d'Hermite et de Stieltjes*. 2 vols. Eds. B. Baillaud and H. Bourget. Paris: Gauthier-Villars.

Herodotus. *The Histories*. Translation 1954 by Aubrey de Sélincourt. Harmondsworth, Middlesex, U.K.: Penguin Books Ltd.

Hsieh, David A. 1988. The statistical properties of daily foreign exchange rates: 1974-1983. *Journal of International Economics* 24: 129-145.

Hurst, H.E. 1946. Instrument-making in Egypt. *Journal of Scientific Instruments* 23: 134-134.

Hurst, H.E. 1951. Long-term storage capacity of reservoirs. *Transactions of the American Society of Civil Engineers* 116: 770-799, 800-808.

Hurst, H.E. 1954. Measurement and utilization of the water resources of the Nile Basin. *Proceedings of the Institution of Civil Engineers* 3 (Part 3): 1-26.

Hurst, H.E. 1956. Methods of using long-term storage in reservoirs. *Proceedings of the Institution of Civil Engineers* 5 (Part I): 519-590.

James, Jessica. 2002. The probability of large daily moves in FX. *CitiFX Investor Strategy 17:* November.

James, Jessica. 2003. Trend following and option writing—a surprising portfolio. *CitiFX Risk Advisory* March.

James, Jessica. 2004. *Currency Management: Overlay and Alpha Trading*. London: Risk Books.

James, Jessica and Hetty Colchester. 2003. Defining forex option value. *Risk* 16 (1).

Jegadeesh, N. and S. Titman. 1993. Returns to buying winners and selling losers: Implications for stock market efficiency. *Journal of Finance* 48: 65-91.

Jovanovic, Franck. 2000. L'origine de la théorie financière: une réévaluation de l'apport de Louis Bachelier. *Revue d'Économie Politique* 110 (3): 395-418.

Jovanovic, Franck and Philippe Le Gall. 2001. Does God practice a random walk? The 'financial physics' of a nineteenth-century forerunner, Jules Regnault. *European Journal of the History of Economics* 8 (3): 332-362.

Kahneman, Daniel and Mark W. Riepe. 1998. Aspects of investor psychology: Beliefs, preferences and biases investment advisors should know about. *Journal of Portfolio Management* Summer: 52-65.

Kettell, Brian. 2001. *Financial Economics.* London: Financial Times Prentice Hall.

Keynes, John Maynard. 1936. *The General Theory of Employment, Interest and Money.* New York: Harcourt, Brace & World.

Kho, Bong-Chan, Dong Lee and René M. Stulz. 2000. U.S. banks, crises and bailouts: From Mexico to LTCM. *American Economic Review* 90 (2): 28-31.

Knauf, Stephane. 2003. Making money from FX volatility. *Quantitative Finance* 3 (3): C48-C51.

Koedijk, Kees. G., Marcia M.A. Schafgans, and Caspar G. de Vries. 1990. The tail index of exchange rate returns. *Journal of International Economics* 29: 93-108.

Lantsman, Y., J.A. Major, and J.J. Mangano. 2002. On the multifractal distribution of insured property. *Fractals* 10 (3): 305-311.

Lo, Andrew W. and A.C. MacKinley. 1988. Stock markets do not follow random walks: Evidence from a simple specification test. *Review of Financial Studies* 1: 41-66.

Lo, Andrew W., ed. 1997. *Market Efficiency: Stock Market Behavior in Theory and Practice*, Vols. I & II. Cheltenham, Glos., U.K.: Edward Elgar Publishing Ltd.

Madan, Dilip B., Peter Carr, and Eric C. Chang. 1998. The variance gamma process and option pricing. *European Finance Review* 2: 79-105.

Maillet, Bertrand and Thierry Michel. 2003. An index of market shocks based on multiscale analysis. *Quantitative Finance* 3: 88-97.

Malkiel, Burton G. 1973. *A Random Walk Down Wall Street.* New York: W.W. Norton & Co.

Mandelbrot, Benoit B. 1951. Adaptation d'un message sur la ligne de transmission, I & II. *Comptes Rendus* (Paris) 232 : 1638-1640, 2003-2005. • Reprint: Mandelbrot 1997b.

Mandelbrot, Benoit B. 1953. Contribution à la théorie mathématique des jeux de communication. *Publications de l'Institut de Statistique de l'Université de Paris* 2: 1-124.

Mandelbrot, Benoit B. 1959. Variables et processus stochastiques de Pareto-Lévy et la répartition des revenus, I & II. *Comptes Rendus* (Paris) 249: 613-615, 2153-2155. • Reprint: Mandelbrot 1997b.

Mandelbrot, Benoit B. 1960. The Pareto-Lévy law and the distribution of income. *International Economic Review* 1: 79-106. • Reprint: Mandelbrot 1997a.• Reprint: *Vilfredo Pareto: Critical Assessments.* Edited by John C. Wood and Michael McLure. London: Routledge, 1999, IV, 155-182. •

Reprint: *Income Distribution,* Edited by Michael Sattinger. *The International Library of Critical Writings in Economics.* Series Editor: Mark Blaug. Edward Elgar, Cheltenham, UK 2000.

Mandelbrot, Benoit B. 1961. Stable Paretian random functions and the multiplicative variation of income. *Econometrica* 29: 517-543. • Reprint: Chapter E11 of Mandelbrot 1997a. • Reprint: *Vilfredo Pareto: Critical Assessments.* Edited by John C. Wood and Michael McLure. London: Routledge, 1999, IV, 183-209.

Mandelbrot, Benoit B. 1962a. Paretian distributions and income maximization. *Quarterly Journal of Economics* 76: 57-85. • Reprint: Chapter E12 of Mandelbrot 1997a • Reprint: *Vilfredo Pareto: Critical Assessments.* Edited by John C. Wood and Michael McLure. London: Routledge, 1999, IV, 210-240.

Mandelbrot, Benoit B. 1962b. Sur certains prix spéculatifs: faits empiriques et modèle basé sur les processus stables additives de Paul Lévy. *Comptes Rendus* (Paris) 254 : 3968-3970. • Reprint: Mandelbrot 1997b, 2004b.

Mandelbrot, Benoit B. 1962c. The variation of certain speculative prices. *IBM Research Report NC-87,* March.

Mandelbrot, Benoit B. 1963a. The variation of certain speculative prices. *Journal of Business* 36: 394-419. • Photographic reprint followed by discussions by Eugene F. Fama and Paul H. Cootner: Cootner 1964, 297-337. • Addendum: Correction of an error in "The variation of certain speculative prices." *Journal of Business*: 45, 1972, 542-543. • Photographic reprint: *Futures Markets.* Three volumes edited by A. G. Malliaris. Cheltenham, UK: Edward Elgar, 1996. 2, 173-198. • Reprint: Chapter E14 of Mandelbrot 1997a. • Reprint: *Classic Futures: Lessons from the Past for the Electronic Age.* Edited by Lester Telser. London: Risk Books. 2000, 649-683.

Mandelbrot, Benoit B. 1963b. The stable Paretian income distribution, when the apparent exponent is near two. *International Economic Review* 4: 111-115.

Mandelbrot, Benoit B. 1963c. New methods in statistical economics. *Journal of Political Economy* 71, 421-440. • Reprint: *Bulletin of the International Statistical Institute, 34th Session, Ottawa* 40 (book 2), 1964, 699-720. • Reprint: Chapter E3 Mandelbrot 1997a. • Reprint: *Vilfredo Pareto: Critical Assessments.* Edited by John C. Wood and Michael McLure. London: Routledge, 1999, IV, 241-263. • Reprint: *Forecasting Financial Markets.* Edited by Terence C. Mills. *The International Library of Critical Writings in Economics.* Series Editor: Mark Blaug. Cheltenham, UK: Edward Elgar, 2002.

Mandelbrot, Benoit B. 1964. Random walks, fire damage amount, and other

Paretian risk phenomena. *Operations Research* 12: 582-585. • Reprint: End of chapter E8 of Mandelbrot 1997a.

Mandelbrot, Benoit B. 1965a. Une classe de processus stochastiques homothétiques à soi; application à la loi climatologique de H.E. Hurst. *Comptes Rendus* (Paris) 260: 3274-7. • Reprints: Mandelbrot 1997b, 2004b. • Translation: Chapter H9 of Mandelbrot 2002.

Mandelbrot, Benoit B. 1965b. Self-similar error clusters in communications systems and the concept of conditional stationarity. *IEEE Transactions on Communications Technology* COM-13: 71-90. • Reprint: Chapter H8 of Mandelbrot 2002.

Mandelbrot, Benoit B. 1966a. Forecasts of future prices, unbiased markets, and "martingale models." *Journal of Business* 39: 242-255. • Reprint: Chapter E19 of Mandelbrot 1997a. • Reprint: *Forecasting Financial Markets*. Edited by Terence C. Mills. *The International Library of Critical Writings in Economics*. Series Editor: Mark Blaug. Cheltenham, UK: Edward Elgar, 2002.

Mandelbrot, Benoit B. 1966b. Nouveaux modèles de la variation des prix (cycles lents et changements instantanés). *Cahiers du Séminaire d'Econométrie* 9: 53-66.

Mandelbrot, Benoit B. 1967a. How long is the coast of Britain? Statistical self-similarity and fractional dimension. *Science* 156: 636-638.

Mandelbrot, Benoit B. 1967b. The variation of some other speculative prices. *Journal of Business* 40: 393-413. • Reprint: Chapter E15 of Mandelbrot 1997a.

Mandelbrot, Benoit B. 1968. Some aspects of the random-walk model of stock market prices: Comment. *International Economic Review* 9: 258-259.

Mandelbrot, Benoit B. 1969. Long-run linearity, locally Gaussian processes, H-spectra and infinite variances. *International Economic Review* 10: 82-111. • Abstract: Intermittency and periodicity, and the problem of long cycles. *Econometrica* 34, 1966 (Supplement): 152-153.

Mandelbrot, Benoit B. 1970. Long-run interdependence in price records and other economic time series. *Econometrica* 38: 122-123.

Mandelbrot, Benoit B. 1972. Possible refinement of the lognormal hypothesis concerning the distribution of energy dissipation in intermittent turbulence. *Statistical Models and Turbulence*. M. Rosenblatt and C. Van Atta, eds. Lecture Notes in Physics 12. New York: Springer, 333-351. • Reprint: Chapter N14 of Mandelbrot 1999a.

Mandelbrot, Benoit B. 1974a. Intermittent turbulence in self-similar cascades; divergence of high moments and dimension of the carrier. *Journal of Fluid Mechanics* 62: 331-358. • Reprint: Chapter N15 of Mandelbrot 1999a.

Mandelbrot, Benoit B. 1974b. Multiplications aléatoires itérées et distributions invariantes par moyenne pondérée aléatoire, I & II. *Comptes Rendus*

(Paris) : 278A ; 289-292 et 355-358. • Reprint : Chapter N16 of Mandelbrot 1999a.

Mandelbrot, Benoit B. 1975. *Les objets fractals : forme, hasard et dimension.* Paris : Flammarion.

Mandelbrot, Benoit B. 1982. *The Fractal Geometry of Nature.* New York: W.H. Freeman & Co.

Mandelbrot, Benoit B. 1985. Self-affine fractals and fractal dimension. *Physica Scripta* 32: 257-260. • Reprint: *Dynamics of Fractal Surfaces.* Edited by Fereydoon Family & Tamas Vicsek. Singapore: World Scientific, 1991, 11-20. • Reprint Chapter H21 of Mandelbrot 2002.

Mandelbrot, Benoit B. 1986. Self-affine fractal sets, I: The basic fractal dimensions, II: Length and area measurements, III: Hausdorff dimension anomalies and their implications. *Fractals in Physics.* Edited by Luciano Pietronero & Erio Tosatti, Amsterdam: North-Holland, 3-28. • Reprint of Part I: *Dynamics of Fractal Surfaces.* Edited by Fereydoon Family & Tamas Vicsek. Singapore: World Scientific, 1991, 21-36. • Reprint in Chapters H22, H23, H24 of Mandelbrot 2002.

Mandelbrot, Benoit B. 1990. Limit lognormal multifractal measures. *Frontiers of Physics: Landau Memorial Conference* (Tel Aviv, 1988). Edited by E. A. Gotsman et al. New York: Pergamon, 309-340.

Mandelbrot, Benoit B. 1997a. *Fractals and Scaling in Finance: Discontinuity, Concentration, Risk.* New York: Springer-Verlag.

Mandelbrot, Benoit B. 1997b. *Fractales, hasard et finance.* Paris: Flammarion.

Mandelbrot, Benoit B. 1997c. Three fractal models in finance: Discontinuity, concentration, risk. *Economic Notes (Banca Monte dei Paschi di Siena SpA)* 26 (2): 197-212.

Mandelbrot, Benoit B. 1997d. Les fractales et la bourse. *Pour la Science* 242: 16-17.

Mandelbrot, Benoit B. 1999a. *Multifractals & 1/f Noise: Wild Self-Affinity in Physics.* New York: Springer-Verlag.

Mandelbrot, Benoit B. 1999b. Renormalization and fixed points in finance, since 1962. *Physica A* 263: 477-487.

Mandelbrot, Benoit B. 1999c. A fractal walk down Wall Street. *Scientific American* February: 70-73.

Mandelbrot, Benoit B. 1999d. Survey of multifractality in finance. *Cowles Foundation Discussion Paper* 1238.

Mandelbrot, Benoit B. 1999e. Randonnées multifractales à Wall Street. *Les mathématiques sociales.* Paris: Belin, 126-130.

Mandelbrot, Benoit B. 2000. Multifractal structure of financial prices and its implications. *Cowles Foundation Paper* 991.

Mandelbrot, Benoit B. 2001a. Scaling in financial prices, I: Tails and dependence. *Quantitative Finance* 1: 113-123. • Reprint: *Beyond Efficiency and Equilibrium*. Edited by Doyne Farmer & John Geanakoplos, Oxford UK: The University Press, 2004.

Mandelbrot, Benoit B. 2001b. Scaling in financial prices, II: Multifractals and the star equation. *Quantitative Finance* 1: 124-130. • Reprint: *Beyond Efficiency and Equilibrium*. Edited by Doyne Farmer and John Geanakoplos. Oxford, UK: The University Press, 2004.

Mandelbrot, Benoit B. 2001c. Scaling in financial prices, III: Cartoon Brownian motions in multifractal time. *Quantitative Finance* 1: 427-440.

Mandelbrot, Benoit B. 2001d. Scaling in financial prices, IV: Multifractal concentration. *Quantitative Finance* 1: 641-649.

Mandelbrot, Benoit B. 2001e. Stochastic volatility, power-laws and long memory. *Quantitative Finance* 1: 558-559.

Mandelbrot, Benoit B. 2002. *Gaussian Self-Affinity and Fractals: Globality, the Earth, 1/f Noise, and R/S*. New York: Springer Verlag.

Mandelbrot, Benoit B. 2003. Heavy tails in finance for independent or multi-fractal price increments. *Handbook on Heavy Tailed Distributions in Finance*. Edited by Svetlozar T. Rachev (*Handbooks in Finance*: 30, Senior Editor: William T. Ziemba): 1, 1-34. • Related paper: *Journal of Statistical Physics* 110, 2003, 739-777.

Mandelbrot, Benoit B. 2004a. *Fractals and Chaos: The Mandelbrot Set and Beyond*. New York: Springer Verlag.

Mandelbrot, Benoit B. 2004b. Updated reprint of Mandelbrot 1997a.

Mandelbrot, Benoit B., Adlai Fisher, and Laurent Calvet. 1997. A multifrac-tal model of asset returns. *Cowles Foundation Discussion Paper* 1164. • See also under Calvet and Fisher.

Mandelbrot, Benoit B. and H.M. Taylor. 1967. On the distribution of stock price differences. *Operations Research* 15: 1057-1062. • Reprint: Chapter E21 of Mandelbrot 1997a.

Mandelbrot, Benoit B. and J.W. Van Ness. 1968. Fractional Brownian motions, fractional noises and applications. *SIAM Review* 10: 422-437. • Reprint: Chapter H11 of Mandelbrot 2002.

Mandelbrot, Benoit B. and James R. Wallis 1968. Noah, Joseph and opera-tional hydrology. *Water Resources Research* 4: 909-918. • Reprint: Chapter H10 of Mandelbrot 2002.

Mandelbrot, Benoit B. and James R. Wallis. 1969a. Computer experiments with fractional Gaussian noises. *Water Resources Research* 5: 228-267. • Reprint: Chapters 12, 13, and 14 of Mandelbrot 2002.

Mandelbrot, Benoit B. and James R. Wallis. 1969b. Some long-run proper-ties of geophysical records. *Water Resources Research* 5: 321-340. • Reprint: Chapter H27 of Mandelbrot 2002.

Mandelbrot, Benoit B. and James R. Wallis. 1969c. Robustness of the rescaled range R/S in the measurement of noncyclic long-run statistical dependence. *Water Resources Research* 5: 967-988. • Reprint: Chapter H25 of Mandelbrot 2002.

Marcus, Alan J. 1990. The Magellan Fund and market efficiency. *Journal of Portfolio Management* Fall: 85-88.

Markowitz, Harry M. 1959. *Portfolio Selection: Efficient Diversification of Investments.* New Haven, CT: Yale University Press.

Markowitz, Harry M. 1990a. Autobiography. Nobel e-Museum, www. nobel.se/economics/laureates/1990/

Markowitz, Harry M. 1990b. Foundations of Portfolio Theory: Nobel Lecture. *Economic Sciences*: 279-287.

Markowitz, Harry M. 1999. The early history of portfolio theory: 1600-1960. *Financial Analysts Journal* 55 (4): 5-16.

Masters, Roger D. 1999. *Fortune is a River. Leonardo da Vinci and Niccolo Machiavelli's Magnificent Dream to Change the Course of Florentine History.* New York: Plume.

McFarland, James W., R. Richardson Pettit, and Sam K. Sung. 1982. The distribution of foreign exchange price changes: Trading day effects and risk measurement. *Journal of Finance* 37 (3): 693-715.

Medova, Elena. 2000. Measuring risk by extreme values. *Risk* November 2000: S20-S26.

Mehra, Rajnish and Edward C. Prescott. 2003. The equity premium in retrospect. *NBER Working Paper 9525.* Cambridge, MA: National Bureau of Economic Research.

Merton, Robert C. 1995. Fischer Black. *Journal of Finance* 50 (5): 1359-1370.

Merton, Robert C. 1997. Autobiography. Nobel e-Museum, www.nobel.se /economics/laureates/1997/.

Miller, Merton H. 1999. The history of finance: An eyewitness account. *Journal of Portfolio Management* Summer: 95-101.

Mills, Terence C. 1993. Is there long-term memory in UK stock returns? *Applied Financial Economics* 3: 303-306.

Mittnik, Stefan, Svetlozar T. Rachev, and Marc S. Paolella. 1998. Stable Paretian modelling in finance: Some empirical and theoretical aspects. Adler et al 1998.

Monopolies and Mergers Commission. 1997. Northern Ireland Electricity PLC: A report on a reference under Article 15 of the Electricity (Northern Ireland) Order 1992.

Müller, U.A., M.M Dacorogna, R.D. Davé, O.V. Pictet, R.B. Olsen, and J.R. Ward. 1993. Fractals and intrinsic time—a challenge to econometricians.

49th International Conference of the Applied Econometrics Association, Luxembourg.

Mulligan, Robert F. 2000. A fractal analysis of foreign exchange markets. *International Advances in Economic Research* 6 (1): 33-49.

Nawrocki, David. 1995. R/S analysis and long term dependence in stock market indices. *Managerial Finance* 21(7): 78-91.

New York State Consumer Protection Board. 2001. In the matter of the Commission as to the rates, charges, rules and regulations of Central Hudson Gas & Electric Corporation for electric and gas service. *Reply Brief of the New York State Consumer Protection Board.*

Officer, R. R. 1972. The distribution of stock returns. *Journal of the American Statistical Association* 67: 807-812.

Pandey, G., S. Lovejoy, and D. Schertzer. 1998. Multifractal analysis of daily river flows including extremes for basins of five to two million square kilometers, one day to 75 years. *Journal of Hydrology* 208: 62-81.

Pareto, Vilfredo. 1896. *Cours d'économie politique.* Reprinted in *Oeuvres Complètes*, 1966, Vol.I. Geneva: Librairie Droz.

Pareto, Vilfredo. 1909. *Manuel d'économie politique.* Traduction française sur l'édition italienne par Alfred Bonnet (revue par l'auteur). Paris : Marcel Giard & Brière. Reprinted in *Oeuvres Complètes*, 1966, Vol VII. Geneva: Librairie Droz.

Patel, Navroz. 2001. Econophysics—does it work? *Risk* March 2001: 33-34.

Peitgen, Heinz-Otto and Dietmar Saupe. 1988. *The Science of Fractal Images.* New York: Springer.

Peters, Edgar E. 1996. *Chaos and Order in the Capital Markets: A New View of Cycles, Prices and Market Volatility.* 2nd Edition. New York: John Wiley & Sons.

Regnault, Jules. 1863. *Calcul des chances et philosophie de la Bourse.* Paris: Mallet-Bachelier.

Richards, Gordon R. 2000. The fractal structure of exchange rates: measurement and forecasting. *Journal of International Financial Markets, Institutions and Money* 10: 163-180.

Rogers, L.C.G. 1997. Arbitrage with fractional Brownian motion. *Mathematical Finance* 7 (1): 95-105.

Roll, Richard. 1970. *The Behavior of Interest Rates: An Application of the Efficient Market Model to U.S. Treasury Bills.* New York: Basic Books.

Rubinstein, Mark. 2002. Markowitz's "Portfolio Selection": A fifty-year retrospective. *Journal of Finance* 57 (3): 1041-1045.

Samuels, Warren J. 1974. *Pareto on Policy.* Amsterdam: Elsevier Scientific Publishing Co.

Samuelson, Paul A. 1970. The fundamental approximation theorem of portfolio analysis in terms of means, variances and higher moments. *Review of Economic Studies* 37 (112) October: 537-542.

Scholes, Myron S. 1995, Fisher Black. *Journal of Finance* 50 (5): 1359–1370.

Scholes, Myron S. 1997. Autobiography. Nobel e-Museum, www.nobel.se/economics/laureates/1997/.

Scholes, Myron S. 2000. Crisis and risk management. *American Economic Review* 90 (2): 17-21.

Scholes, Myron S. 2001. Merton H. Miller: Memories of a great mentor and leader. *Journal of Finance* 56 (4): 1179-1182.

Schoutens, Wim. 2003. *Lévy Processes in Finance: Pricing Financial Derivatives.* Chichester, West Surrey, UK: John Wiley & Sons Ltd.

Securities and Exchange Commission. 1998. *Trading Analysis of October 27 and 28, 1997: A Report by the Division of Market Regulation of the U.S. Securities and Exchange Commission.* On the Web at http://www.sec.gov/news/studies/tradrep.htm.

Shahin, Mamdouh. 1985. *Hydrology of the Nile Basin.* Amsterdam: Elsevier Science Publishers BV.

Shalizi, Cosma. 2003. The world is our laboratory. *Journal of Quantitative Finance* 3 (2): C20-21.

Sharpe, William F. 1964. Capital asset prices: A theory of market equilibrium under conditions of risk. *Journal of Finance* 19 (3): 425-442.

Sharpe, William F. 1990a. Autobiography. Nobel e-Museum, www.nobel.se/economics/laureates/1990/.

Sharpe, William F. 1990b. Capital asset prices with and without negative holdings: Nobel Lecture. *Economic Sciences*: 312-332.

Shiller, Robert J. 2000. *Irrational Exuberance.* Princeton, NJ: Princeton University Press.

Schwert, G. William. 2004. http://schwert.simon.rochester.edu/volatility.htm

Simons, Katerina. 2000. The use of Value at Risk by institutional investors. *New England Economic Review* November/December: 21-30.

Singh, Vijay P. and David A. Woolhiser. 2002. Mathematical modeling of watershed hydrology. *Journal of Hydrologic Engineering* 7 (4): 170-292.

Sprott, Julien Clinton. 2003. *Chaos and Time-Series Analysis.* Oxford, UK: Oxford University Press.

Sullivan, Edward J. and Timothy M. Weithers. 1991. Louis Bachelier: the father of modern option pricing theory. *Journal of Economic Education* 22 (2): 165-171.

Summers, Lawrence H. 2000. International financial crises: causes, prevention, and cures. *American Economic Review* 90 (2): 1-16.

Taqqu, Murad S. 2001. Bachelier and his times: a conversation with Bernard Bru. *Finance and Stochastics* 5 (1): 3-32.

Teichmoeller, John. 1971. A note on the distribution of stock price changes. *Journal of the American Statistical Association* 66 (334): 282-284.

Thind, Sarfraz. 2002. Fewer options in 2001: A Risk survey. *Risk* 15 (5).

Valdez-Cepeda, Ricardo David, and Enrique Solano-Herrera. 1999. Self-affinity of records of financial indexes. *Fractals* 7 (4): 427-432.

Von Koch, Helge. 1905. Une méthode géométrique élémentaire pour l'étude de certaines questions de la théorie des courbes planes. *Acta Mathematica* 30: 145-174.

Walter, Christian. 2001. Searching for scaling laws in distributional properties of price variations: A review over 40 years. *Colloquium Report, International Actuarial Association*, Toronto, Canada, 7 September.

Walter, Christian. 2002. 1900-2000: Un siècle de descriptions statistiques des fluctuations boursières, ou, les aléas du modèle de marche au hasard. *Colloque Marché Boursier*. Paris: Collège de France.

Watsham, Terry J. and Keith Parramore. 1997. *Quantitative Methods in Finance*. London: Thomson Learning.

Weibel, E.R. 1963. *Morphometry of the Human Lung*. New York: Academic.

Weron, Rafal and Beata Przybylowicz. 2000. Hurst analysis of electricity price dynamics. *Physica A* 283: 462-468.

Willinger, Walter, Murad S. Taqqu, and Vadim Teverovsky. 1999. Stock market prices and long-range dependence. *Finance and Stochastics* 3: 1-13.

Zumbach, G., M. Dacarogna, J. Olsen, and R. Olsen. 2000. Measuring shocks in financial markets. *International Journal of Theoretical and Applied Finance* 3: 347-55.

Index

Enron
    risk seen in, 8
Equity Premium Puzzle, 230–231
Euclid, 34, 123, 129
Eureka moment, 147
Exceptional chance
    law of, 159–162
    Lévy and, 159–160
    probability distributions of, 162
Executive stock options
    Bachelier's theories influencing, 14
Exogenous effects
    cat's brains with, 241
    economics with, 228
Extreme Value Theory, 273

Fama, Eugene F., 11, 55, 75, 96
    α estimates calculated by, 262
    Mandelbrot introduced by, 166, 167
    market dependence studied by, 99
    paper by, 102, 104, 289
    portfolio risk suggested by, 265–266
FASB. *See* Financial Accounting
        Standards Board
Fat tails, 168, 170–172
    α with, 262
    Bouchaud's model with, 260
    fractal properties in, 198, 208
    Nile flooding with, 174
    research need for, 274
Federal Reserve Board, 107, 219, 275
Feller, Willy, 31, 32, 35
Fever charts, 17
FIGARCH, 15
Filter method, 235–236, 300
Finance
    analyzing investments in, 261–264
    Black-Scholes influence on, 70–76,
        167
    Brownian motion price assumption
        of, 86–87
    CAPM in, 59–60
    case against modern theory of, 79–107
    continuous price change assumption
        of, 85–86, 235–238
    fractals in, 261
    global crises in, 274–275
    investors all alike assumption of,
        84–85
    managing risk in, 271–274
    Markowitz's influence on, 61–66, 167
    mistakes of, 80–82
    modern, 9, 59–77, 167
    portfolios building in, 265–267
    primitive state of, 262
    rational people assumption of, 83–84
    shaky assumptions of, 82–87
    Sharpe's influence on, 66–70

    ten heresies of, 225–252
    value concept in, 249–252
    valuing options in, 268–271
Financial Accounting Standards Board
        (FASB)
    Internet bubble and, 280
    valuing options regulations by, 271
Financial Analysts Journal, 285
Financial market
    black box view of, 28
    chance in, 27–30, 41–42
    clusters in, 216
    coin-tossing view of, 49–54
    deceptiveness of, 244–247
    dependence in, 206
    flexibility of time in, 238–240
    fractal geometry with, 5–6
    improbable events influencing, 3–4
    memory in, 12
    mild path in, 42
    multifractal model for, 207–222, 239
    orthodox theory of, 46
    parable of, 242–243
    pictorial essay of, 88–94
    price changes in, 208, 235–238
    probability in, 26
    risk in, 4
    roughness in, 123–131
    trend following in, 82
    turbulence in, 114–117
    volatility in, 234
Financial modeling
    alien visitor in, 198–200
    Bachelier in, 10, 117–120
    Brownian, 92–94
    dependence in, 98–100, 182, 184–185,
        186, 188
    development of, 198
    fractal cartoon in, 117–122, 171
    fractional Brownian motion in, 19,
        188, 191–192
    importance of, 197
    irregular trends observed in, 199
    multifractal model in, 22, 207–222
    Oanda as, 254
    random walk model in, 9, 12, 18, 21
    volatile volatility in, 19
Fisher, Adlai, 217
Foreign exchange
    dependence in, 191
    modern finance limitations for, 97–98
    multifractal modeling of, 217–219
    Oanda experiment with, 253–259
    valuing options for, 268
Forex houses
    Oanda like, 253, 256
    technical analysis employed by, 9
Fourier, Jean Baptiste Joseph, 53